Current Topics in Neuroendocrinology Volume 4

This collection of studies was conceived as part of a two volume review of the subject. The contents of Volume 6 are listed below.

Neurobiology of Oxytocin

Neurobiology of Vasopressin

Editors
D. Ganten and D. Pfaff

Contributors
G. Clarke R.E. Lang M.J. McKinley L.P. Merrick
W. Rascher D. Richter M. Sofroniew Th. Unger
A. Weindl

With 53 Figures

Springer-Verlag
Berlin Heidelberg New York Tokyo

Editors

Dr. DETLEV GANTEN, M.D., Ph.D.
Pharmakologisches Institut
Universität Heidelberg
Im Neuenheimer Feld 366
6900 Heidelberg/FRG

Dr. DONALD PFAFF, Ph.D.
Rockefeller University
York Avenue, and 66th Street
New York, NY 10021/USA

The picture on the cover has been taken from Nieuwenhuys R., Voogd J., van Huijzen Chr.: The Human Central Nervous System. 2nd Edition. Springer-Verlag Berlin Heidelberg New York 1981

ISBN-13:978-3-642-68495-1 e-ISBN-13:978-3-642-68493-7
DOI: 10.1007/978-3-642-68493-7

Library of Congress Cataloging in Publication Data. Main entry under title: Neurobiology of vasopressin. (Current topics in neuroendocrinology; v. 4) Includes bibliographies and index. 1. Vasopressin. 2. Neurobiology. I. Ganten, D. (Detlev), 1941–. II. Pfaff, Donald W., 1939–. III. Clarke, G. IV. Series. [DNLM: 1. Vasopressins–physiology. W1 CU82Q v. 4/WK 520 N4936] QP572.V3N48 1985 599'.0188 85-4797
ISBN-13:978-3-642-68495-1 (U.S.)

2121/3130-543210

Contents

Biosynthesis of Vasopressin

D. RICHTER[1]

Contents

1 Historical Account

Since the first reports by van Dyke and co-workers (reviewed in Acher 1979) on the isolation of a protein complex then called the "van Dyke protein" and later identified as being composed of the hormone vasopressin and its corresponding carrier protein, known as neurophysin numerous publications have appeared dealing with the synthesis of vasopressin in the hypothalamus and its axonal transport to the neurohypophysis (reviewed in Sachs 1969; Valtin et al. 1974; Breslow 1979; Acher 1979; Brownstein et al. 1980; Pickering 1983).

In 1964 Sachs and Takabatake presented evidence that the nonapeptide vasopressin might be synthesized on ribosomes via a biologically inactive precursor. A few years later, Sachs proposed a remarkable hypothesis for the biosynthetic pathway of this hormone (Sachs et al. 1969):

According to this model (precursor model) the biosynthesis of the peptide bonds in vasopressin would occur solely in the perikaryon on ribosomes, via pathways common to

1 Institut für Physiologische Chemie. Abteilung Zellbiochemie, Universität Hamburg, Martinistr. 52, D-2000 Hamburg 20

Current Topics in Neuroendocrinology, Vol. 4

the biosynthesis of other peptide chains; initially however, the hormones would be constructed as part of a macromolecule (probably a protein). The release of the octapeptide from the precursor molecule presumably takes place during the formation and maturation of the NSG (neurosecretory granules). It is unknown whether or not neurophysin biosynthesis occurs via a similar mechanism or whether the hormone and neurophysin share a common precursor. The biosynthesis of vasopressin and neurophysin nevertheless appear closely related.

It was this pioneering concept of Sachs which finally led to the structure of the vasopressin and oxytocin precursors (Land et al. 1982, 1983). The early attempts to identify the proposed vasopressin precursor were principally based on pulse-chase experiments carried out in vivo in normal and mutant rat strains (Brattleboro rats); the latter lack biologically active vasopressin and hence provide a model for studying the coordinate synthesis of vasopressin and its respective carrier protein, neurophysin (reviewed in Valtin et al. 1974).

With refined techniques Brownstein, Gainer, and associates (Brownstein et al. 1980) succeeded in the isolation of high-molecular-weight proteins from rat hypothalami which were composed of vasopressin- or oxytocin-like peptides and their corresponding neurophysins.

The final resolution of the structure of the vasopressin precursor was achieved by a combination of cell-free translation studies and recombinant DNA techniques from which the amino acid sequence could be deduced (reviewed in Richter 1983 a). These data proved unambiguously that the vasopressin precursor is composed of several distinct peptides, the hormone, its carrier protein neurophysin, and a glycoprotein.

This article concentrates on the attempts leading to the elucidation of the biosynthetic pathway and the structure of the vasopressin precursor. Since oxytocin is structurally similar to vasopressin, overlapping data are included where relevant; in a separate section the structures of the two hormone precursors are compared.

1.1 Analysis by Pulse-Chase Experiments

The early work by Sachs and co-workers showed that ^{35}S-labeled vasopressin was synthesized in the hypothalamus after injection of ^{35}S-cysteine. Subsequent pulse-chase experiments indicated that the hormone synthesis took place on ribosomes, since it could be inhibited by the antibiotic puromycin (Sachs and Takabatake 1964). On the other hand, biologically active vasopressin was not associated with the ribosomal fraction; this led to the suggestion that vasopressin was initially synthesized as an inactive precursor on ribosomes, and later converted into the mature hormone at a site different from the ribosomes.

Extending his studies on the biosynthesis of the neurophysins, Sachs then presented evidence that synthesis of vasopressin and its respective carrier protein, neurophysin, may be correlated (Sachs 1969). This assumption was based on the observations that vasopressin and its neurophysin were found in the same subcellular fractions, and that neurosecretory granules contained both peptides, which are released simultaneously from organ cultures of the neurohypophysis.

1.2 The Neurophysin Precursors

Direct application of ^{35}S-cysteine to the supraoptic nucleus of the brain, sampling of the tissue by the micropunch technique, and subsequent analysis by electrofocusing led to the identification of precursors to the neurophysins (Brownstein et al. 1980). One precursor could be converted by limited trypsinization into neurophysin and an oligopeptide which, although not identical with vasopressin, showed features similar to vasopressin as indicated by immunoreactivity and affinity chromatography; the other precursor gave rise to oxytocin-associated neurophysin and an oxytocin-like peptide.

Although these data suggested the existence of a composite precursor, as proposed by Sachs, this model was still debated. It was not clear why the potential precursors failed to cross-react with antibodies against vasopressin or oxytocin. Possibly, the failure of the immunological identification was due to the vasopressin or oxytocin antibodies used, which discriminated between acidic and amidated forms of the hormones, a specificity of the antibodies unsuitable for detecting potential hormone precursors.

Although partial trypsinization of the precursor gave rise to neurophysin it was not possible to identify authentic vasopressin$_{1-8}$, which ought to have been released if the hormone was present. This discrepancy was probably due to the limited trypsinization conditions applied in these experiments.

Extraction of hypothalamic tissue and identification by specific antibodies revealed a glycosylated precursor to neurophysin and vasopressin, suggesting that this prohormone is a glycoprotein (Lauber et al. 1979). Further studies by Cohen and associates (Beguin et al. 1981; Lauber et al. 1981) indicated the existence of a protein even larger than the vasopressin prohormone, with immunologic determinants for vasopressin, neurophysin, ACTH, and β-endorphin. This group proposed a model whereby the larger composite precursor serves as "mother ship", for all the other hormones referred to. Direct sequence analysis is needed to understand the nature of this precursor.

1.3 Biosynthesis and Structure of the Hormone Precursor

The complete structural organization of the postulated vasopressin precursor became evident by a combination of cell-free translation studies and recombinant DNA technology. The experimental concept adopted included (a) translation of hypothalamic mRNA and immunological identification of the products; (b) tryptic peptide mapping; and (c) cloning and sequencing of the cDNA encoding the hormone.

1.3.1 Translation of Hypothalamic mRNA

Cell-free translation systems in general offer a simple way to study the immediate translation product of an exogenous mRNA. As with other hormones, vasopressin is synthesized as a primary translate, the preprohormone which contains a so-called signal peptide at its NH_2 terminus, an essential marker for transportation into the lumen of the endoplasmic reticulum (Kreil 1981). The signal peptide is cleaved off during synthesis, a process which can be simulated in a cell-free system by complementation with membranes prepared from the endoplasmic reticulum.

The in vitro biosynthesis of vasopressin or oxytocin was studied in cell-free systems programmed with hypothalamic mRNA from mouse, rat, or calf (reviewed in Chaiken 1983; Gainer 1983; Richter 1983 a). The cell-free systems used were wheat-germ extracts, rabbit reticulocyte lysates, or *Xenopus laevis oocytes*, with ^{35}S-cysteine or tritiated amino acids as radioactive markers. The translation products were identified by immunoprecipitation followed by electrophoresis in dodecyl sulfate on polyacrylamide gels, and visualized by autoradiography.

The cell-free synthesized products were initially identified by immunological means. By this method precursors with apparent molecular weights of 21000–25000 and 16000–19000 were obtained with antisera raised against the respective neurophysins.

In order to identify antigenic determinants for vasopressin or oxytocin within these precursors, specific antisera were applied which did not discriminate between the acid or deamidated form of the hormones; an essential prerequisite in the search for potential vasopressin or oxytocin precursors. By means of sequential immunopecipitation it became evident that the initially identified precursors also contained antigenic determinants for vasopressin and oxytocin: the bovine

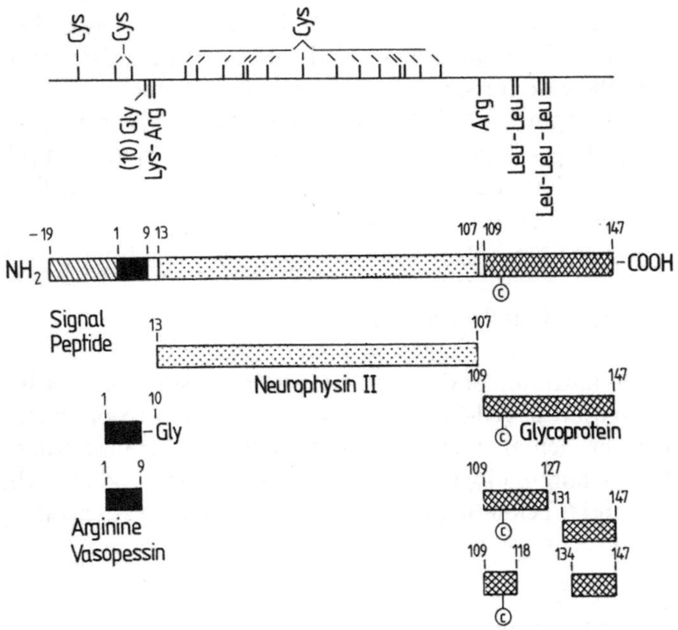

Fig. 1. Scheme of the calf vasopressin precursor. *Hatched bar* represents signal peptide; the *solid bars*, vasopressin; *Stippled bars*, neurophysin; *cross-hatched bars*, glycoprotein and its fragments; *open bars*, amino acids involved in processing. The arrangement of cysteine residues, glycine residue for amidation, the basic amino acids lysine and arginine, and leucine residues as signals for proteolysis is shown at the top of the figure; *C*, carbohydrate chain

```
                                            -23
                                            MET LEU ALA MET MET LEU ASN THR THR LEU SER ALA CYS
RAT        AGCAGAGCAGAGCTGCACGCA-GTGCCCACCT ATG CTC GCC ATG ATG CTC AAC ACT ACG CTC TCT GCT TGC
CALF GCACAGTA********CA****G*A*GT******---- *C* *GT *** *G* *** *C* G** G*C **A **G C*C **C ***
                                                            ——— PRO ASP ALA ——————— PRO ———————

                                    -1  +1          VASOPRESSIN
PHE LEU SER LEU LEU ALA LEU THR SER ALA CYS TYR PHE GLN ASN CYS PRO ARG GLY     GLY LYS
TTC CTG AGC CTG CTG GCC CTC ACC TCT GCC TGC TAC TTC CAG AAC TGC CCA AGA GGA     GGC AAG
*** **C *** *** *** *** T** *** *** **T *** *** *** *** *** *** *** **G **C     *** ***
——————————————————————— PHE ———————————————————————————————————————————————     ———————

            NEUROPHYSIN             20
ARG     ALA THR SER ASP MET GLU LEU ARG GLN CYS LEU PRO CYS GLY PRO GLY GLY LYS GLY ARG
AGG     GCC ACA TCC GAC ATG GAG CTG AGA CAG TGT CTC CCC TGC GGC CCT GGC GGC AAA GGG CGC
***     *** *TG *** *** C** *** *** *** *** *** *** *** *** *** **C **G *** *** **C ***
———     ——— MET ——————— LEU ———————————————————————————————————————————————————————————

                            40
CYS PHE GLY PRO SER ILE CYS CYS ALA ASP GLU LEU GLY CYS PHE LEU GLY THR ALA GLU ALA LEU
TGC TTC GGG CCG AGC ATC TGC TGC GCG GAC GAG CTG GGC TGC TTC CTG GGC ACC GCC GAG GCG CTG
*** *** *** **C *** *** *** *** *G* *** *** *** *** *** *** G** *** **G *** *** *** ***
——————————————————————————————— GLY ——————————————————————— VAL ———————————————————————

                    60
ARG CYS GLN GLU GLU ASN TYR LEU PRO SER PRO CYS GLN SER GLY GLN LYS PRO CYS GLY SER GLY
CGC TGC CAG GAG GAG AAC TAC CTG CCC TCG CCC TGC CAG TCT GGC CAG AAG CCT TGC GGA AGC GGA
*** *** **A *** *** *** *** *** **G *** *** *** *** **C *** *** *** **C *** **G *** **G
———————————————————————————————————————————————————————————————————————————————————————

            80
GLY ARG CYS ALA ALA ALA GLY ILE CYS CYS SER ASP GLU SER CYS VAL ALA GLU PRO GLU CYS ARG
GGC CGC TGC GCT GCC GCG GGC ATC TGC TGC AGC GAT GAG AGC TGC GTG GCC GAG CCC GAG TGT CGA
*** *** *** **C *** **C *** *** *** *** *A* *** *** *** *** *** A** *** *** *** **C **G
——————————————————————————————————————— ASN ——————————————————— THR ———————————————————

                                                    GLYCOPROTEIN
GLU GLY PHE PHE ARG LEU THR             ARG     ALA ARG GLU GLN SER ASN ALA THR GLN LEU
GAG GGT TTT TTC CGC CTC ACC --- ---     CGC     GCT CGG GAG CAG AGC AAC GCC ACG CAG CTG
**A *** G*C GG* TT* *C* CG* CGC GTT     ***     **C AAC **C *G* *** *** **G **C *T* ***
——————— VAL GLY PHE PRO ARG ARG VAL     ———     ——— ASN ASP ARG ——————————————— LEU ———

            120
ASP GLY PRO ALA ARG GLU LEU LEU LEU ARG LEU VAL GLN LEU ALA GLY THR GLN GLU SER VAL ASP
GAC GGG CCA GCC CGG GAG CTG CTG CTT AGG CTG GTA CAG CTG GCT GGG ACA CAA GAG TCC GTG GAT
*** *** **G AG* G** *CC T** T** **G C** *** **G *** *** **G *** G*G *CG *** C** *C* **G
——————————— SER GLY ALA ——————————————————————————————————————— ALA PRO ——— PRO ALA GLU

    140
SER ALA LYS PRO ARG VAL TYR STOP
TCT GCC AAG CCC CGG GTC TAC TGA  GCCATGC-------CCCCCCACGCCTCCCCCCTACAGCATGGAAAATAAAC-TT
C*C *** C** *** G*C *** *** ***  *G*GCGCCCCCCC*T******C***G-***TGG*****C*A*********G**
PRO ——— GLN ——— GLY ———————————


TTAAAAAA(POLY A)
*****GGC(POLY A)

              3'
```

Fig. 2. Comparison of the nucleotide and amino acid sequences of calf and rat vasopressin precursors. The *positive numbers* indicate the positions of the amino acids; the *negative numbers*, amino acids of the signal sequence. Identical nucleotides in both sequences are shown by *asterisks*, identical amino acids by *solid lines; broken lines* indicate a lack of corresponding sequences of either precursor; absence of an amino acid is shown by a gap

21 000-dalton preprohormone was composed of vasopressin and neurophysin II, the other, with a molecular weight of 16 500, of oxytocin and neurophysin I (Schmale and Richter 1981 a).

Cell-free translation of mRNA extracted from either supraoptic or paraventricular hypothalamic nuclei of salt-treated rats indicated that vasopressin precursor synthesis was considerably increased with mRNA derived from the supraoptic nuclei (Richter 1983 b). Whether this result reflects a regulatory mechanism of vasopressin expression in different hypothalamic regions remains to be studied in greater detail.

1.3.2 Tryptic Peptide Mapping

Based on two-dimensional chromatography of the peptides obtained after trypsinization of the vasopressin precursor, a single unambiguous model for the order of the involved peptides of bovine preprovasopressin was proposed. In this model, the signal sequence was followed by the hormone, the respective neurophysin, and a glycoprotein at the C terminus (Schmale and Richter 1981 b). Experiments with rat provasopressin cleaved by cyanogen bromide predicted a similar alignment (Russell et al. 1981).

1.3.3 Cloning and cDNA Sequencing

Conclusive proof of the intramolecular organization of the composite precursors, as well as of the nature of the glycoprotein, was obtained by sequence analysis of cloned cDNA encoding the bovine vasopressin precursor. The sequence confirmed the predicted organization of the vasopressin precursor.

The bovine preprovasopressin was found to consist of 166 amino acids and to have a molecular weight of 17310; the rat vasopressin precursor which could be predicted from the sequence of the vasopressin gene was almost similar in size, with 168 amino acid residues and a molecular weight of 17826 (Figs. 1 and 2). The molecular weights obtained by SDS-polyacrylamide gel electrophoresis of the in vitro-synthesized translation products from rat (19000 daltons) and calf (21000 daltons) seemed to give falsely high estimates; this could be due to the abnormal behavior of the neurophysins in gels.

1.4 Relevance of the Precursor Structure

The primary structure of the vasopressin precursor offered a number of interesting insights into the organization of the composite precursor. First, the hormone was directly adjacent to the signal peptide; consequently, vasopressin is N-terminally located in the prohormone. The sizes of the signal peptides – 19 and 23 amino acid residues for the bovine and rat precursors respectively – are in good agreement with the results of the membrane-supplemented translation experiments (Schmale and Richter 1981). The use of alanine as the signal sequence cleavage site in both precursors conformed with over 50% of other known signal peptidase cleavage sites.

Second, vasopressin and its corresponding neurophysin were separated by the amino acids glycine, lysine, and arginine. Whereas glycine is thought to serve as nitrogen donor for the amidation of the hormone, the basic amino acid residues function as signals for converting the prohormone into the mature peptides. Since the elucidation of the structure of the melittin precursor (Suchanek and Kreil 1977), the glycine residue has been found in all precursors to amidated oligopep-

Table 1. Partial amino acid sequences of precursors to oligopeptides amidated at the carboxy terminus (Richter 1983 a)

Melittin	Gln	Gly	COOH	
α-MSH	Val	Gly	Lys	Lys
γ-MSH	Phe	Gly	Arg	Arg
Vasopressin	Gly	Gly	Lys	Arg
Oxytocin	Gly	Gly	Lys	Arg
Caerulein	Phe	Gly	Arg	Arg
Gastrin	Phe	Gly	Arg	Arg
Calcitonin	Pro	Gly	Lys	Lys
Human joining peptide	Glu	Gly	Lys	Arg
Egglaying hormone	Lys	Gly	Lys	Arg
CRF	Ala	Gly	Lys	

tides (see Table 1). This pattern is complemented by studies using a pituitary enzyme that catalyzes the oxidative transamidation of a model tripeptide only when a glycine residue occupies the C terminus (Bradbury et al. 1982).

Within the precursors the basic amino acids C-terminal to the glycine can either be lacking (melittin) or appear as single (CRF) or dibasic residues (e.g., vasopressin). In the latter case the combination lysine-arginine is prefered over arginine-arginine or lysine-lysine; the combination arginine-lysine has not yet been found. It appears that the removal of the basic amino acids is a prerequisite for amidation (Bradbury et al. 1982).

Third, neurophysin and the glycoprotein within the precursor were separated only by a single basic amino acid, an arginine residue. This may suggest that the converting signals can be either single basic amino acids or pairs. Specific enzymes could discriminate between the two possible cleavage sites. Interestingly, the basic amino acids within the region of the C terminus of bovine neurophysin are not cleaved by the processing enzymes; this suggests that the potential cleavage site is exposed and better accessible to proteolytic enzyme(s) than are the two basic amino acids in the immediate neighborhood.

Finally, the sequence of the glycoprotein consisting of 39 amino acids agreed precisely with that of a glycopeptide isolated from a calf neurohypophysis. Comparison of the calf sequence with those from rat, pig, and man showed remarkable homologies (Holwerda 1972; Smyth and Massey 1979; Seidah et al. 1981). The glycosylation site Asn-Ala-Thr and the leucine-rich center part are well conserved in all species so far known. The consecutive leucine residues may represent alternative processing signals for converting the glycoprotein into subfractions (Smyth and Massey 1979). Although the glycoprotein has been localized in vasopressin-producing magnocellular neurons (Watson et al. 1982), its biological function remains obscure.

2 Structural Organization of the Gene

2.1 Structure of the Rat Vasopressin Gene

According to restriction mapping and nucleotide sequence analysis, the rat vasopressin gene is approximately 2 kilobase pairs long (Schmale et al. 1983). As with other genes from higher eukaryotes, the coding regions (exons A–C) were found to be separated by two intervening sequences (introns I and II) with the principal functional domains on three distinct exons (Figs. 3 and 4).

Exon A contained the untranslated region, the sequence for the signal peptide, the hormone, and the variable part of the N terminus of the neurophysin. Exon B encoded the highly conserved center part of the neurophysin, exon C the variable C terminus of the neurophysin, the glycoprotein, and the untranslated 3′ region of the precursor.

The structural organization of the vasopressin gene was in line with those of other split genes and included the consensus sequences at exon-intron junctions, a modified "Goldberg-Hogness" sequence, CATAAAT located 29 nucleotides

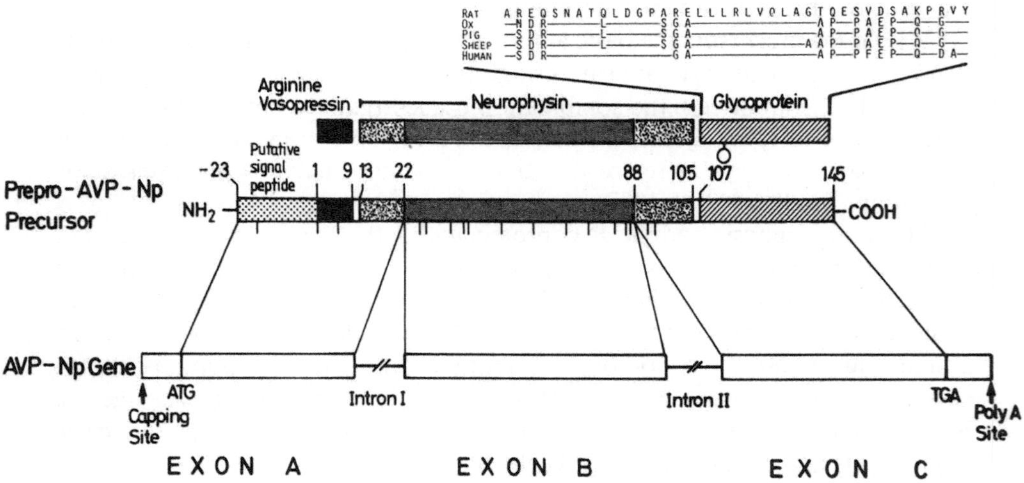

Fig. 3. Scheme of the rat vasopressin gene. *Stippled bar*, putatative signal sequence; solid bars, vasopressin; *shaded bar*, neurophysin (*dark shading* shows the conserved region); *hatched bar*, glycoprotein; *open bars*, post-translational processing signals; ♁, carbohydrate chain. Significant amino acids are *numbered*, the *negative number* refers to the signal sequences; cysteine residues are indicated by *vertical lines*. The box shows the amino acid sequences of glycoproteins from various species

```
-367 CCTGCTAGTCCTTGGTGAATGAGACCTGGGGACCCCTCTAGTCTGTTGAGAGCTGCTGAAATGCTCA

-300 ACTATGATTTCCAGGTGACCCTCAAGTCGGCTCACCTCCCTGATTGCACAGCACCAATCACTGTGGCGGTGGCTCCCGTCACACGGTGGCCAGTGACAGC

-200 CTGATGGCTGGCTCCCCTCCTCCACCACCCTCTGCATTGACAGGCCCACGTGTGTCCCCAGATGCCTGAATCACTGCTGACAGCTTGGGACCTGTCAGCT

-100 GTGGGCTCCTGGGGAGCCACTGGGGAGGGGGTTAGCAGCCACGCTGTCGCCTCCTAGCCAACACCTGCAGACATAAATAGACAGCCCAGCCCGCTCAGGC

                                     MetLeuAlaMetMetLeuAsnThrThrLeuSerAlaCysPheLeuSerLeuLeuAlaLeuThrSerAla
   1 AGCAGAGCAGAGCTGCACGCAGTGCCCACCTATGCTCGCCATGATGCTCAACACTACGCTCTCTGCTTGCTTCCTGAGCCTGCTGGCCCTCACCTCTGCC

       V A S O P R E S S I N                    N P         exon A       intron I
     CysTyrPheGlnAsnCysProArgGlyGlyLysArgAlaThrSerAspMetGluLeuArgGln
 101 TGCTACTTCCAGAACTGCCCAAGAGGAGGCAAGAGGGCCACATCCGACATGGAGCTGAGACAGGTACCACTGTGGTCCGTTCAGGGCTGCTGACAGTGCC

 201 GTAGGAAGGGTCATGGGCTAGGAGAGAGGGAAACCTTGTCTGAGCAGTCAGACTTTAGGGGAGGTTCCTGGAAGGAAGCAGTTATCTTATATGGAGTAGA

 301 TGGGTTTCCCAGAACGGTAAGAGGGGACCAGGTGCCAGAGAAGCCACATAAAGGACAGTGTCCCCAGGCAGGGGATATGCCAGAAAATGAGAGATACTTA

 401 TCACTGGGCTTGGGATGAGAACGGGTTAAACTGGGTACCCTGGCCTCCTCTGCACAGCTGGAGGTGGCCGGTGGTATGTTGGCTCACCAGGACTGGGTAG

 501 ATGGTACGAAACTGTTCTCGCCTGAGTACAAAGCCTTTCCCACCCAGCTCAAACTCTCTTAGCTCCTTTTTTAGCCAGCTGCACCGGTTTCTTCCTGTCC

 601 ACGGAAGACGGCCATTGCCCTGTGTCTGAGCGGAGTATGTCCCACATCTAGCCTCAGCCTCGTGCCCAGATCTGCTGTACTGTATGTTCAGCTCTGAGTC

 701 TGCCCTTCCGGCAGGGCTGAAGGGAATCCAGTCACTAGGCTCAAATCTGGTCAGGTCACAGGTGGCTCAGTTTTGAACAAGCTCGATGGGCAGTAGGCAG

 801 TTCACCGAGTCTGCCTTCCGTTTGCTGAGTTCCTTTGGAGACTTCCGAGGCACTAGGTGTGTCTTGCACCCATCAGCCTAATTCGGTCCTTGCCACCTTC

 901 CTACTAGGGCATAATAGGTTGGCGGGAGGTAAAAGCCCACCAGCGTGGGGCAGGGGTAAGAGTGAGCGAGCCGTAGGTACAGGAAAGAGGATCTTGGAAT

1001 GTGTAGGGCCATCTGAATGTCGGAGAGGTAAGTCTCTGAGAGACTGCTGCACACCGGTGACACATCAGAGCTGAGGAGGTCCCCCAAGTGTTGTCTCCCC

1101 CGCCCCCCGCCCCATACGACTCTGTCAAAGCAGGAGAGGGTTTTGAGACCTCATGAGAACTGATCCTCCTGATAACCTAGCCGGTTAGATTTCCACTCTC

1201 GCCCTTTACGGCTGCTTCGTCCTAGATAGAGCCAGAGCATCTGGCCGGTGAAGCTGGGATAGCAGCAGGGTGACCTTAGGTTCCCAACGCCCCTCTTGGC

                   intron I     exon B
                               CysLeuProCysGlyProGlyGlyLysGlyArgCysPheGlyProSerIleCysCysAlaAspGluL
1301 CTGGCTCCAGCTGACCCGCGTCCTTCCCCGCAGTGTCTCCCCTGCGGCCCTGGCGGCAAAGGGCGCTGCTTCGGGCCGAGCATCTGCTGCGCGGACGAGC

     euGlyCysPheLeuGlyThrAlaGluAlaLeuArgCysGlnGluGluAsnTyrLeuProSerProCysGlnSerGlyGlnLysProCysGlySerGlyGl
1401 TGGGCTGCTTCCTGGGCACCGCCGAGGCGCTGCGCTGCCAGGAGGAGAACTACCTGCCCTCGCCCTGCCAGTCTGGCCAGAAGCCTTGCGGAAGCGGAGG

                    exon B      intron II
     yArgCysAlaAlaAlaGlyIleCysCysSerAspG
1501 CCGCTGCGCTGCCGCGGGCATCTGCTGCAGCGATGGTGCGCACAAAGCCAGGCGGGCTGAGCATGGGGAATGGATGGGGTGGGTGGGAGGTAAAGGGGGG

1601 CTAAGTGGGGGACTGAGGAATCAGGACCGGAGATGGAGGGTGAGTAGTATGAAGGGGGTCGAGAGTTGGAACGTAGCAGGGTAGGATAAAGGGGATTGTG

                                                                         intron II       exon C
                                                                        luSerCysValAlaGluProGluCysArgGluGlyP
1701 GGGATGGCGCCCCTATAGGTGCGCCCACCCCAGGACGCCTGACCTCACACAGCCCTTCCTTCAGAGAGCTGCGTGGCCGAGCCCGAGTGTCGAGAGGGTT

          N P                 G L Y C O P R O T E I N
     hePheArgLeuThrArgAlaArgGluGlnSerAsnAlaThrGlnLeuAspGlyProAlaArgGluLeuLeuLeuArgLeuValGlnLeuAlaGlyThrGl
1801 TTTTCCGCCTCACCCGCGCTCGGGAGCAGAGCAACGCCACGCAGCTGGACGGCCAGCCCGGGAGCTGCTGCTTAGGCTGGTACAGCTGGCTGGGACACA

     nGluSerValAspSerAlaLysProArgValTyrEnd
1901 AGAGTCCGTGGATTCTGCCAAGCCCCGGGTCTACTGAGCCATCGCCCCCCACGCCTCCCCCCTACAGCATGGAAAATAAACTTTTAAAAAA
```

Fig. 4. Sequence of the rat vasopressin gene. The rat anti-sense strand is shown together with the inferred amino acid sequence. Numbers indicate the nucleotide sequence upstream *(negative numbers)* and downstream *(positive numbers)* from the transcription start site. The *underlined region* shows the Goldberg-Hogness sequence. The sequence shown differs from that reported previously (Schmale et al. 1983; Schmale and Richter 1984) at five positions: −95 to −91 reads CTCCT (instead of CTCT); −85 to −81 reads GCCAC (instead of GCAC); −57 (C instead of G); +1 550 to +1 554 reads AGGCG (instead of AGCGC); +1 733 (G instead of A). Amino acid +143 reads Arg (instead of Ala)

upstream from the presumptive transcription start site, an A residue bounded by a primidine-rich region at the transcription start site, and a polyadenylation site, AATAAA, at the 3′ end of the sequence (Breathnach and Chambon 1981).

2.2 Vasopressin Deficiency in Brattleboro Rats

Animals with hereditary hypothalamic diabetes insipidus lack biologically active vasopressin and its corresponding neurophysin, while oxytocin and its carrier are not affected. Characteristically, water resorption is greatly impaired in the distal kidney tubules (reviewed in Valtin et al. 1974).

Isolation of the vasopressin gene from the mutant rat strain has indicated no differences in the restriction cleavage sites when compared with the wildtype gene. Sequence analysis of the Brattleboro vasopressin gene showed that the mutated gene lacks a single nucleotide in the second exon. The mutation leads to an open reading frame predicting a precursor with a different C terminus. The mutated gene is transcribed. No translation products could be identified yet (Schmale and Richter 1984).

3 Expression in Heterologous Systems

Since the functional hormone is synthesized as a longer precursor which is physiologically inactive, its conversion to the modified end product is of critical importance in understanding the regulation of its expression. While the first processing steps can be monitored in cell-free systems by the cotranslation addition of microsomal membranes, such processing steps as those that take place in the Golgi vesicles, or during packaging and transport within neurosecretory vesicles, cannot be easily studied in in vitro system. The *Xenopus laevis* oocyte system may be of some advantage in this respect, since in vivo processing events including secretion can be followed under quasi-in vitro conditions.

3.1 In Vitro Conversion and Glycosylation

Complementation of a reticulocyte lysate system with microsomal membranes from dog pancreas provided the first insights into the biosynthetic pathway of preprovasopressin (Richter 1983a). In this system, the bovine preprohormone (21000 daltons) was converted into the core-glycosylated provasopressin (23000 daltons).

Several experiments consistently indicated the glycoprotein nature of the vasopressin prohormone (23000 daltons): it was bound specifically to concanavalin-A sepharose; it was sensitive to treatment with α-mannosidase, which converted it from a 23000-dalton into a 19000-dalton product; and it could be labeled with radioactive sugars in oocytes programmed with injected hypothalamic mRNA (Richter 1983a).

No carbohydrate chain was found in the oxytocin prohormone. Whether the glycosylation of provasopressin plays a role in specifying post-translational modification steps, such as preventing rapid degradation of the precursor or exposing or masking significant proteolytic cleavage sites, can only be speculated at present.

3.2 Synthesis and Conversion in an Oocyte System

For the purpose of studying further steps in the processing of the vasopressin precursor the oocyte from *Xenopus laevis* was programmed with hypothalamic mRNA. The major vasopressin precursor synthesized within the oocyte corresponded to the cell-free synthesized core-glycosylated prohormone. Another product found in the oocyte was an intermediary form which consisted only of vasopressin and neurophysin (Fig. 5, left part). This product and a more glycosylated prohormone were both secreted by the oocyte into the incubation medium. Neither mature vasopressin nor neurophysin could be detected in the oocyte or

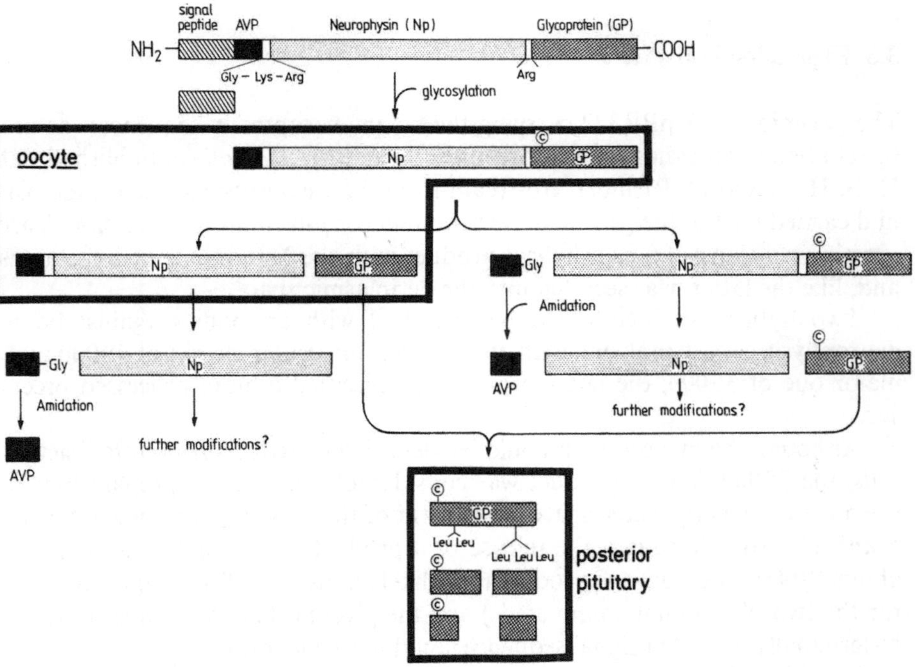

Fig. 5. Possible processing pathways of the vasopressin precursor. The *left framed* part shows processing steps observed in oocytes programmed with hypothalamic mRNA. Alternative processing steps are indicated on the right. Processed fragments of the glycoprotein have been found in the *posterior pituitary*. (Smyth and Massey 1979)

in the incubation medium. The experiments suggested that the converting enzyme(s) of the oocyte recognized the single arginine residue as a cleavage signal separating neurophysin and the glycoprotein – not, however, the pair of basic amino acids between vasopressin and neurophysin. Different sets of converting enzymes are probably involved in the maturation process of the hormone. The release of vasopressin from the precursor or the intermediary form may be closely linked to the amidation of its C terminus – a process which may require more tissue-specific enzyme(s).

The elucidated structure of the vasopressin precursor implied that for the release of the functional peptides in vivo, specific processing must take place to remove the signal peptide, to separate the C-terminal glycoprotein, and to cleave the hormone from its neurophysin. Except for the release of the hormone, all these stages, as well as the secretion from the cell, appear to be correctly executed by the frog oocyte. Whether the sequence of processing events observed in the oocyte also takes place in the secretory vesicles during the axonal transport to the neurohypophysis is not known yet. The processing model proposed in Fig. 5 shows the events so far observed in the oocyte (left part); a vasopressin-neurophysin intermediate form may also be present in the neurohypophysis (Masse et al. 1982). Alternatively, the first cleavage could occur between vasopressin and neurophysin; a neurophysin-glycoprotein could then be further processed into the respective peptides (Fig. 5, right part).

3.3 Expression in Bacteria

The plasmid DNA pBR322 carrying the bovine vasopressin sequence was used for studying expression of this precursor in *Escherichia coli* (unpublished data H. G. Hertl and D. Richter). Upstream from the vasopressin sequence the plasmid carried part of the gene and the promoter for β-lactamase expression. Vasopressin was translated as a hybrid product with the N terminus of β-lactamase and, like the latter, was secreted into the periplasmic space.

Two hybrid products could be identified with antibodies against bovine neurophysin II, a minor one with an apparent molecular weight of 38 000 and a major one of 35 000; the latter probably represented a more processed precursor.

Although the two products could be identified in extracts from total bacterial cells, the 35 000-daltons product was considerably enriched in preparations derived from the periplasmic space. Treatment of the 35 000-dalton precursor with cyanogen bromide caused the release of a product with a molecular weight of about 18 000. The cleavage product most likely consisted of neurophysin (except for the two N-terminal amino acids) and the glycoprotein. Vasopressin with its N-terminally extended signal sequence could not be identified.

The identification of the hybrid protein in the periplasmic space of *E. coli* showed that it was able to adopt a conformation that hindered neither its translocation into the periplasmic space nor the correct folding of the neurophysin antigenic determinants.

4 Comparison of Vasopressin and Oxytocin Precursor Structures

As had been anticipated on the basis of the in vivo and in vitro studies, the oxytocin precursor did not contain a glycoprotein at the carboxy terminus; instead,

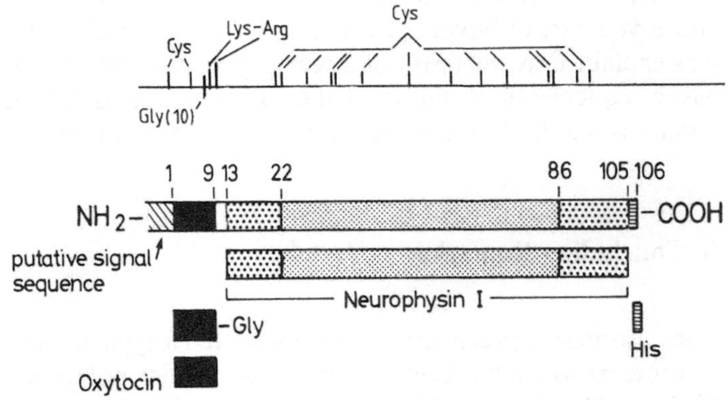

Fig. 6. Scheme of the bovine oxytocin precursor. *His*, histidine residue; *hatched bar*, signal peptide; *solid bars*, oxytocin; coarsely stippled, variable N and C terminus of NpI; finely stippled, conserved sequence of NpI

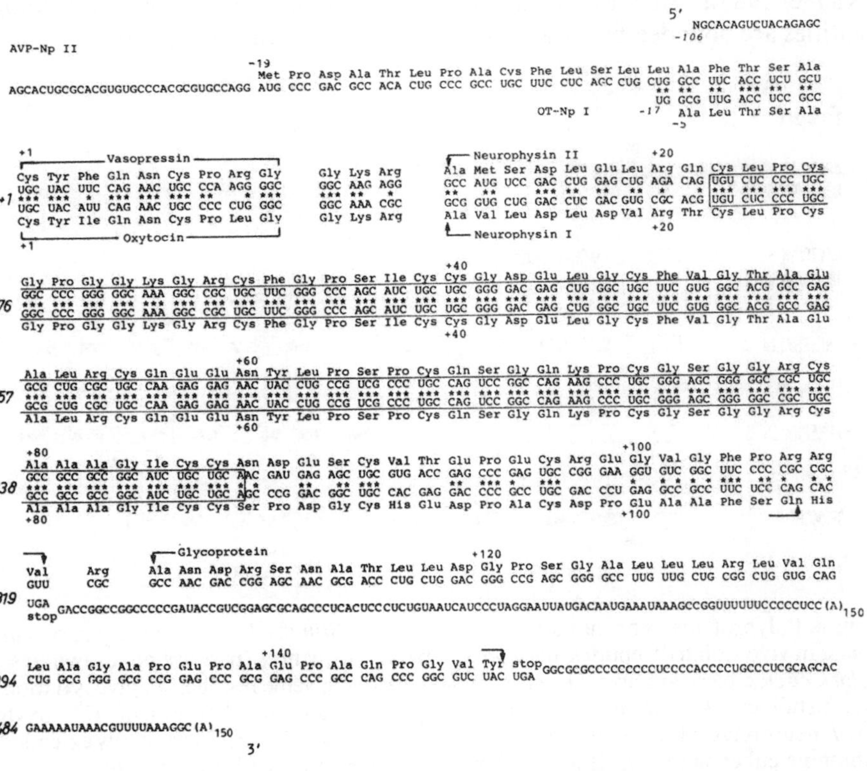

Fig. 7. Nucleotide and amino acid sequence of the bovine oxytocin and vasopressin precursors. Symbols are as in Fig. 1

the C terminus of the neurophysin was extended by one extra amino acid residue, histidine, presumably cleaved off during post-translational processing (Fig. 6). In the corresponding rat sequence this extra amino acid is an arginine residue (Ivell and Richter 1984).

Comparison of the nucleotide sequences of the vasopressin and oxytocin mRNAs revealed a 197-nucleotide-long stretch of homology encoding the highly conserved part of bovine neurophysin I and II (Fig. 7). This sequence homology was explained as the result of a recent gene conversion event (Land et al. 1983). More sequence information of the vasopressin and oxytocin genes from other species is required for further insights into the evolution of the two genes.

5 Concluding Remarks

The vasopressin precursor can be considered a typical representative of the class of proteins which has been referred to as cellular polyproteins (Fig. 8; Koch and Richter 1980; Herbert 1981); this definition has been introduced in analogy to viral polyproteins which are composed of several distinct viral entities. Other examples of cellular polyproteins are the precursors to ACTH, the enkephalins (reviewed in Herbert 1981; Numa and Nakanishi 1981), or oxytocin (reviewed in Richter 1983 a). All these precursors have in common the fact that their biological entities are bounded by basic amino acids used as cleavage signals.

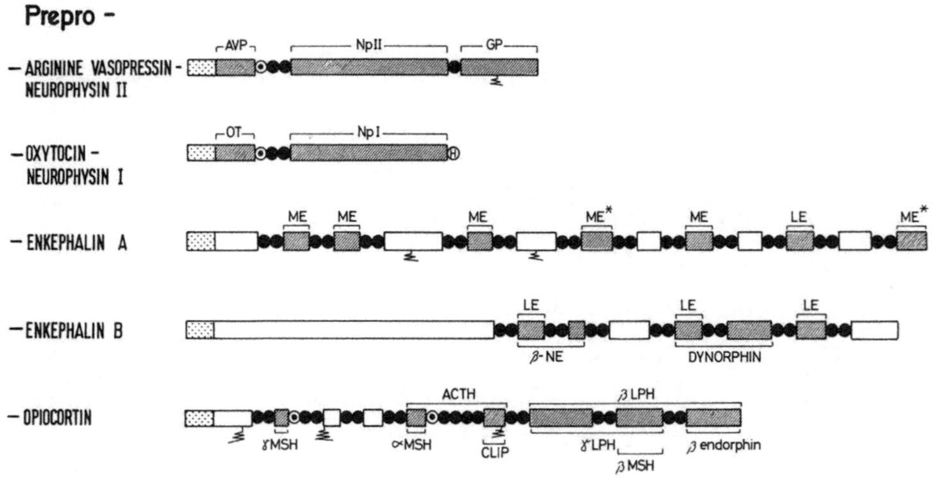

Fig. 8. Polyproteins as precursors to neuropeptides. *Stippled bars*, signal sequence; *hatched bars*, in vivo isolated peptides; *open bars*, "spacer" sequences (bars are not drawn to scale); *black circles*, basic amino acids; *open circles with dot*, glycine residue; ⚸, glycosylation site; *H*, histidine; *AVP*, arginine vasopressin; *NpII*, neurophysin II from calf; *OT*, oxytocin; *NpI*, neurophysin I from calf; *ME*, methionine enkephalin; *ME*, C-terminally extended methionine enkephalin; *LE*, leucine enkephalin; *β-NE*, β-neoendorphin; *β-* or *γ-LPH*, *β-* or *γ*-lipotropin; *MSH*, melanotropin-stimulating hormone; *CLIP*, corticotropin-like intermediate-lobe peptide; *ACTH*, adrenocorticotropin; *GP*, glycoprotein

Hormones such as vasopressin have attracted attention as potential models for studying gene expression and regulation because of their multifunctional properties. Vasopressin, for example, increases blood pressure and controls water resorption (Valtin et al. 1974); it helps to release ACTH (Schally et al. 1978) and is considered to be involved in memory and learning processes (Burbach et al. 1983). Nucleotide sequence analysis has helped considerably to clarify the organization of this hormone precursor; analysis of the genomic DNA encoding the vasopressin gene in Brattleboro rats should provide further insight into the question of how regulation of such multifunctional precursors is monitored.

References

Acher R (1979) Neurophysine – molekulare und zellulaere Aspekte. Angew Chem 91:905–919

Beguin P, Nicolas P, Bousetta H, Fahy C, Cohen P (1981) Characterization of the 80 000 molecular-weight form of neurophysin isolated from bovine neurohypophysis. J Biol Chem 256:9289–9294

Bradbury AF, Finnie MDA, Smyth DG (1982) Mechanism of C-terminal amide formation by pituitary enzymes. Nature 298:686–688

Breathnach R, Chambon P (1981) Organization and expression of eucaryotic split genes coding for proteins. Ann Rev Biochem 50:349–383

Breslow E (1979) Chemistry and biology of the neurophysins. Ann Rev Biochem 48:251–274

Brownstein MJ, Russell JT, Gainer H (1980) Synthesis, transport, and release of posterior pituitary hormones. Science 207:373–378

Burbach JPH, Kovacs GL, Wang X-C, de Wied D (1983) Metabolites of arginine-vasopressin and oxytocin are highly potent neuropeptides in the brain. In: Koch G, Richter D (eds) Biochemical and Clinical Aspects of Neuropeptides. Academic, New York 211–222

Chaiken IM (1983) Neurophysin-neuropeptide hormone complexes: biosynthetic origin and noncovalent interactions. Biopolymers (in press)

Gainer H (1983) Precursors of vasopressin and oxytocin. Prog Brain Res 60:205–215

Herbert E (1981) Discovery of pro-opiomelanocortin – a cellular polyprotein. TIBS 6:184–188

Holwerda DA (1972) A glycopeptide from the posterior lobe of pig pituitaries. Eur J Biochem 28:340–346

Ivell R, Richter D (1984) Structure and comparison of the oxytocin and vasopressin genes from rat. Proc Natl Acad Sci USA 81:2006–2010

Koch G, Richter D (1980) Biosynthesis, modification, and processing of cellular and viral polyproteins. Academic, New York

Kreil G (1981) Transfer of proteins across membranes. Ann Rev Biochem 50:317–348

Land H, Schuetz G, Schmale H, Richter D (1982) Nucleotide sequence of cloned cDNA encoding bovine arginine vasopressin-neurophysin II precursor. Nature 295:299–303

Land H, Grez M, Ruppert S, Schmale H, Rehbein M, Richter D, Schuetz G (1983) Deduced amino acid sequence from the bovine oxytocin-neurophysin I precursor cDNA. Nature 302:342–344

Lauber M, Camier P, Cohen P (1979) Immunological and biochemical characterization of distinct high-molecular-weight forms of neurophysin and somatostatin in mouse hypothalamus extracts. FEBS Lett 97:343–347

Lauber M, Nicolas P, Boussetta H, Fahy C, Beguin P, Camier M, Vaudry H, Cohen P (1981) The Mr 80 000 common forms of neurophysin and vasopressin from bovine neurohypophysis have corticotropin- and endorphin-like sequences and liberate by proteolysis biologically active corticotropin. Proc Natl Acad Sci USA 78:6086–6090

Masse MJO, Desbois-Perichon P, Cohen P (1982) Identification of neurophysin-related proteins in bovine neurose-cretory granules. Eur J Biochem 127:609–617
Numa S, Nakanishi S (1981) Corticotropin-lipotropin precursor – a multi-hormone precursor – and its gene. TIBS 6:274–277
Pickering BT (1982) The neurosecretory neuron in the 1980s. In: Farner DS, Lederis K (eds) Neurosecretion. Plenum, New York, pp 415–430
Richter D (1983 a) Synthesis, processing and gene structure of vasopressin and oxytocin. In: Cohn WE, Holdave K (eds) Progress in Nucleic Acid Research 30. Academic, New York, pp 245–266
Richter D (1983 b) The neuropeptide hormone arginine vasopressin and hypothalamic diabetes insipidus. In: Austin L, Jeffrey P (eds) Molecular Aspects of Neurological Disorders. Academic Sydney, pp 321–329
Richter D, Schmale H, Ivell R, Kalthoff HW (1982) Cellular polyproteins from pituitary and hypothalamus. Composite precursors to oligopeptide hormones. In: McKerns KW, Pantic V (eds) Hormonally Active Brain Peptides. Plenum, New York 581–598
Russell JT, Brownstein MJ, Gainer H (1981) Biosynthesis of neurohypophyseal polypeptides: The order of peptide components in propressophysin and pro-oxyphysin. Neuropeptides 2:59–65
Sachs H (1969) Neurosecretion. Adv Enzymol 32:327–372
Sachs H, Takabatake Y (1964) Evidence for a precursor in vasopressin biosynthesis. Endocrinology 75:943–948
Sachs H, Fawcett P, Takabatake Y, Portanova R (1969) Biosynthesis and release of vasopressin and neurophysin. Recent Prog Horm Res 25:447–491
Schally AV, Coy DH, Meyers CA (1978) Hypothalamic regulatory hormones. Ann Rev Biochem 47:89–128
Schmale H, Richter D (1981 a) Immunological identification of a common precursor to arginine vasopressin and neurophysin II synthesized by in vitro translation of bovine hypothalamic mRNA. Proc Natl Acad Sci USA 78:766–769
Schmale H, Richter D (1981 b) Tryptic release of authentic arginine vasopressin from a composite arginine vasopressin/neurophysin II precursor. Neuropeptides 2:47–52
Schmale H, Heinsohn S, Richter D (1983) Structural organization of the rat gene for the arginine vasopressin-neurophysin precursor. EMBO J 2:763–767
Schmale H, Richter D (1984) Single base deletion in the vasopressin gene is the cause of diabetes insipidus in Brattleboro rats. Nature 308:705–709
Seidah NG, Benjannet S, Chretien M (1981) The missing fragment of the pro-sequence of human pro-opiomelanocortin sequence and evidence for C-terminal amidation. Biochem Biophys Res Commun 100:901–907
Smyth DG, Massey DE (1979) A new glycopeptide in pig, ox and sheep pituitary. Biochem Biophys Res Commun 87:1006–1010
Suchanek G, Kreil G (1977) Translation of melittin mRNA in vitro yields a product terminating with glutaminylglycine rather than with glutaminamide. Proc Natl Acad Sci USA 74:975–978
Valtin H, Stewart J, Sokol HW (1974) Genetic control of the production of posterior pituitary principles. In: Knobil E, Sawyer WH (eds) Handbook of Physiology, Hanover New Hampshire 7:31–171
Watson SJ, Seidah NG, Chretien M (1982) The carboxy terminus of the precursor of vasopressin and neurophysin: immunocytochemistry in rat brain. Science 217:853–855

Electrophysiological Studies of the Magnocellular Neurons

G. CLARKE and L. P. MERRICK [1]

Contents

1 Introduction

The magnocellular neurons of the hypothalamus which synthesize and release the peptide hormones vasopressin and oxytocin present a fascinating and attractive target for the electrophysiologist. These neurons are relatively large and are clustered together in dense aggregates, mostly in the paraventricular nuclei (PVN)

1 Department of Anatomy, University of Bristol, University Walk, Bristol BS8 1TD Great Britain

Current Topics in Neuroendocrinology, Vol. 4

and supraoptic nuclei (SON), which can be easily located. They have relatively simple neuronal profiles and send the vast majority of their axons to the neural lobe of the pituitary, which itself is remote from the rest of the brain and consists predominantly of the terminals of the magnocellular neurons. These features allow the neurons to be recorded and identified electrophysiologically. The ultrastructural and biochemical characteristics of these neurons also allow them to be readily visualized histologically. The neurosecretory product of the magnocellular neurons can be detected and measured in the plasma, and several physiological reflexes cause a selective activation of the neurons, resulting in the release of detectable quantities of hormone. During the past 25 years considerable insight has been gained into the functions of these neurons and the significance of their electrical activity; yet, though information regarding them continues to accumulate, there are many fundamental aspects which remain unclear. Our understanding of their organization and regulation, rather than being enhanced, has actually become more uncertain. Nevertheless, as modern immunohistochemical techniques have developed to reveal other peptide-containing neurons within the brain, the magnocellular neurons of the hypothalamo-neurohypophyseal system have taken on the role of the model system with which other hypothalamic and extrahypothalamic peptidergic neurons are compared.

In this chapter it is our intention to provide a general review of the topic, but not to dwell on details which have been comprehensively covered in many previous articles. Our aim is to consider some of the novel and interesting features which have been observed and the concepts which have developed over the past few years. We shall also concentrate on aspects of regulation of these neurons by various neurotransmitters, a subject which has been rather neglected to date. Most of the data in this chapter refer to the rat, which has been the animal of choice in most investigations and, indeed, has provided some of the more informative facts; reference to other species will be made where relevant. Finally, in order to study the vasopressin neurons it is necessary to consider in detail the related oxytocin neurons; this is done in Sect. 4.5.4.

2 Identification and Characterization of the Magnocellular Neurons

2.1 Antidromic Identification

The location of the majority of the magnocellular neurosecretory neurons in the SON and PVN of the hypothalamus allows microelectrodes to be placed in the vicinity of these neurons by means of stereotaxic techniques. A variety of electrodes have been used, ranging from electrolyte-filled glass micropipettes to several types of metal electrodes; in the case of chronic recordings from unanaesthetized animals these are often just fine platinum wires (25–30 μm in diameter), insulated with varnish except at the exposed end. The hypothalamus is usually entered by either a dorsal approach from the superior surface of the brain or a transpharyngeal approach to expose the mediobasal surface of the hypothalamus. This latter, ventral approach allows the optic tracts to be visualized, and thus the SON to be readily located, but requires extensive surgery with associated blood loss and stress to the animal. The dorsal approach (the method favoured by the authors) requires little surgery but makes stereotaxic placement more arduous, there by reducing the success rate for positioning the electrodes within the magnocellular nuclei. The dorsal approach is used in experiments involving the chronic placement of microwires and the recording of magnocellular neurons from unanaesthetized animals. As the SON and PVN can be identified histologically, it is possible to check for the correct placement of microelectrodes at postmortem. Early experiments had to rely on this retrospective assessment of whether the recordings had been made from magnocellular neurons, and no account was taken of the heterogeneous nature of the nuclei.

Experimental credibility was greatly enhanced when the technique of antidromic identification was introduced into experiments in which neurosecretory neurons were recorded (Yagi et al. 1966). Antidromic identification relies on the principle that when an axon is stimulated an action potential will be propagated retrogradely (antidromically) along the fibre to invade the cell soma. It is essential to this technique of identification that only neurosecretory cells project to the area being stimulated; in the case of the neural lobe the vast majority of axons do indeed originate from magnocellular neurons.

Three criteria are commonly used to characterize an antidromic potential. These are (a) constant latency, (b) high-frequency following, and (c) collision extinction (Fig. 1). These criteria are explained more fully elsewhere (Dreifuss et al. 1981) but will be outlined briefly here. The latency of the antidromic potential will be constant for any one neuron since it is dependent only upon the conduction time of the axon, which will not alter, being a function of the physical properties of the cell. In our experience, supraoptic neurons in the lactating rat anaesthetized with urethane have a mean latency of about 14 ms (range 6–20 ms); similar values have been reported by others (Dyball and McPhaill 1974). In recordings in unanaesthetized rats (Summerlee and Lincoln 1981) somewhat lower values (8 ms) have been observed; this is perhaps not surprising, since an effect of anaesthetics on conduction velocity has been reported (Dyball and McPhaill 1974). The estimated conduction velocity of 0.5–1 m/s for magnocellular neurons corresponds to the velocities expected for C fibres (see Poulain and Wakerley 1982). There is, however, a curious anomaly with regard to the antidromic propagation time which has often been reported, namely a jump to a shorter latency. Thus, in some 50% of the neurons we have recorded there was a shorter latency when the stimulus strength was increased from the threshold level. In the majority of neurons displaying this phenomenon it was a change of 1–3 ms; this is attributed to the stimulus spreading to a more proximal part of the axon, or to a collateral with a faster conduction velocity. The criterion of "high-frequency following" (the ability of the neuron to follow every stimulus pulse with an antidromic potential) is based on the observation that synaptic transmission fails at a frequency lower than that of axonal conduction. Caution is required in the interpretation of such experiments, however, for the fine, unmyelinated axons of the magnocellular neurons are not particularly fail-safe at high frequencies. The most reliable criterion is collision; this makes use of the fact that an action potential travelling retrogradely along the axon will extinguish, and itself be extinguished by, one travelling in an

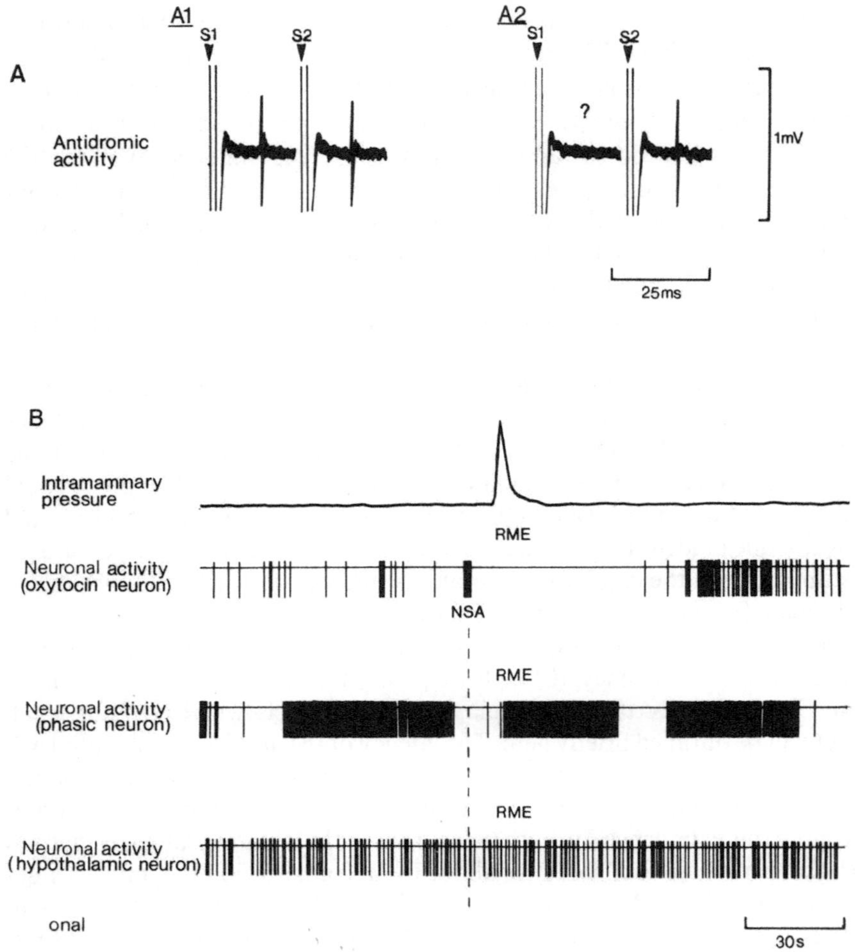

Fig. 1. **a** Oscilloscope records showing the generation of antidromic potentials when the neural lobe was stimulated (*S*). In each case five sweeps were superimposed. The characteristics of constant latency and frequency following are illustrated on the left. Following each stimulus pulse (*S1* and *S2*) an antidromic potential was always generated with a latency of 13 ms; the two stimuli were 20 ms apart (=frequency of 50 Hz). The phenomenon of collision is illustrated on the right. In this case the stimulator was triggered by a spontaneous action potential 1 ms prior to S1; the antidromic potential which should have occurred 13 ms after S1 was extinguished by the spontaneous orthodromic potential travelling down the axon. The second antidromic potential followed S2 as usual. **b** Traces showing the activity of three types of hypothalamic neurons at the time of reflex milk ejection (*RME*). The traces have been positioned to align the moment of reflex milk ejection (*dotted line*); each pen deflection indicates the occurrence of an action potential. The upper trace is the intramammary pressure recorded simultaneously with the oxytocin neuron. The upper trace of neuronal activity was identified as being from an oxytocin neuron on the basis of antidromic identification, and by the occurrence of a short period of high-frequency neurosecretary activity (*NSA*) 12 s before the increase in intramammary pressure. The middle trace shows a neuron, also antidromically identified, which showed no high-frequency activity prior to milk ejection but discharged in a phasic fashion typical of putative vasopressin neurons. The lower trace shows a hypothalamic neuron which could not be stimulated antidromically

orthograde direction (owing to the refractory state of the adjacent membrane). Of course, the validity of each of these techniques is dependent upon the correct placement of the stimulating electrode in the neural lobe rather than in the adjacent median eminence. The correct location can be verified at postmortem; however, it is more convenient if a suitable position can be ensured before recording begins. If a lactating animal is used this can be achieved by simultaneously recording the intramammary pressure; electrical stimulation of the neural lobe using suitable parameters (for the rat, 5-s trains at 50 Hz, of biphasic, 1-ms-square wave pulses of approx 300 μA) will release a pulse of oxytocin, resulting in a rise in intramammary pressure.

2.2 Activity of Magnocellular Neurons

In the unstimulated anaesthetized rat the majority of identified magnocellular neurons either show a slow irregular pattern of activity (< 3 spikes/s) or are completely quiescent and found only by means of antidromic stimulation. This slow activity appears to be a feature of hypothalamic neurons in general (Lincoln 1969) but is not attributable to the anaesthetics used, for similarly low levels of activity are recorded in the unanaesthetized rat (Summerlee and Lincoln 1981). Higher levels of activity can be recorded in magnocellular neurons however, particularly when stimulated in such a way as to release neurohypophyseal hormones (see Sect. 4). When discharging at higher frequencies these neurons routinely show two other patterns of activity – fast continuous and phasic (for definitions, see Poulain et al. 1977). Phasic activity consists of periods of fairly steady discharge (usually in the range of 5–15 spikes/s) lasting several seconds or even minutes, interspersed with periods of complete silence or a few intermittent action potentials (Figs. 1 and 3). Although the activity of the magnocellular neurons is enhanced by stimuli which release the hormones as expected, these neurons do appear to have some intrinsic activity. Thus, in animals in whom the total afferent input has been removed there is still some activity, albeit at a much lower mean rate of discharge (Dyball and Prilusky 1981). Similarly, activity can be recorded from magnocellular neurons in the isolated hypothalamic slice in vitro (Hatton et al. 1978) and even from cultured supraoptic neurons (Gahwiler and Dreifuss 1979).

2.2.1 Classification by Activity

By 1970 the activity of the hypothalamic magnocellular neurons could be correlated with stimuli which released neurohypophyseal hormones, but no further advance could be made until the neurons could be classified as either vasopressin-containing or oxytocin-containing neurons. This major step was achieved electrophysiologically several years before immunohistological techniques made it possible to distinguish visually between the two types of neurons. The method for classifying the magnocellular neurons on the basis of their neuronal activity originated from some observations made by Wakerley and Lincoln in the early 1970s. They first observed that anaesthetized rats released pulses of oxytocin during suckling, and these could be detected by recording the rise in intramammary pressure that each produced (Lincoln et al. 1973). Having noted that this milk ejection reflex operated even under anaesthesia, they were then able to study the activity of the magnocellular neurons during the period in which the pulse of oxytocin was being released (Wakerley and Lincoln 1973). They found that magnocellular neurons of the PVN could be divided into two populations, one group (approximately 60%) that showed a characteristic stereotyped neurosecretory burst of potentials some 15–20 s prior to the rise in intramammary pressure, and a second group whose activity showed no correlation with the milk ejection response (see Fig. 1) Neurons which displayed a high-frequency burst of potentials related to the milk ejection response could also be recorded in the SON, and accounted for about 45% of the antidromically identified neurons (Lincoln and Wakerley 1974). Since the milk ejection reflex in the rat had been shown to result from the selective release of oxytocin (Wakerley et al. 1973), they concluded that those neurons displaying a burst of potentials were the oxytocin neurons. As the neurons which responded always discharged prior to every mammary pressure response, and as the latency from the burst of activity to the pressure response was always similar, they further concluded that all oxytocin neurons respond prior to every milk ejection in a synchronized fashion. It thus became possible to identify oxytocin neurons – at least in the anaesthetized lactating rat.

Unfortunately, there is no stimulus which is as selective for the release of vasopressin as suckling is for the release of oxytocin. However, it is generally agreed that vasopressin neurons can be identified in part by the characteristic activity that they sometimes display, namely "phasic activity". Wakerley and Lincoln (1973) noted that most identified oxytocin neurons showed slow, irregular activity between neurosecretory bursts; some showed fast, continuous activity, but very few discharged with the phasic pattern of activity. Thus, it has to be assumed either that phasic activity is a feature of the vasopressin neurons or that oxytocin neurons showing phasic activity are unresponsive at milk ejection. Subsequent experiments have strengthened the case for phasic activity being the prerogative of the vasopressin neurons (Wakerley et al. 1975; Dreifuss et al. 1976; Poulain et al. 1977), and this is now used as a means for distinguishing between oxytocin and vasopressin neurons in the absence of the milk ejection reflex. It should be noted, however, that the term "phasic activity" encompasses a very broad range of characteristics, and that most investigators have recorded a few neurons showing phasic activity which have also given a burst of potentials in synchrony with the oxytocin neurons at milk ejection; this is a feature which is even more apparent in the unanaesthetized rat (Summerlee 1983). Unfortunately, no one to date has done the definitive experiment of intracellulary labelling phasically firing neurons and then showing that they all exhibit vasopressin-like immunoreactivity. At present, vasopressin neurons can be identified only by default, being those antidromically identified magnocellular neurons which do not respond with a high-frequency burst prior to milk ejection. Such neurons may not be exclusively vasopressin neurons, however (see Sect. 2.3.2)

2.2.2 Oxytocin Neurons and Neurosecretory Activity

Identifiable oxytocin neurons can be recorded in both the SON and the PVN and constitute approximately 50% of the neurons at each site. This is in close agree-

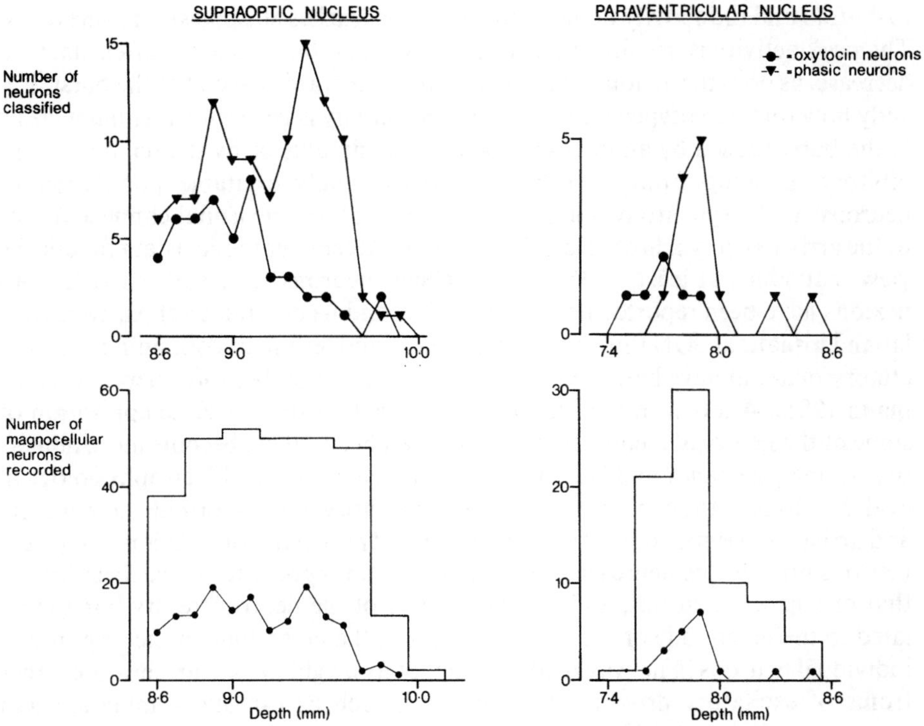

Fig. 2. Histograms showing the vertical distribution of magnocellular neurons recorded in the supraoptic and paraventricular nuclei. The lower histograms show the number of antidromically identified magnocellular neurons encountered in each nucleus in 200-μm steps; the abscissa refers to the depth of the microelectrode tip from the surface of the brain. In some experiments the young were placed on the nipples to evoke the milk-ejection reflex so that the magnocellular neurons could be identified as either oxytocin or nonoxytocin (vasopressin) neurons. The distribution of these identified neurons is shown in the upper graphs and the proportion of the total that they represent is shown by the graphs within the lower histograms. Note that the oxytocin neurons tended to be in the dorsal half of the nuclei

ment with the occurrence of oxytocin neurons identified using immunocytochemical techniques. In a systematic study in which we recorded neurons in the SON we found that the majority of the oxytocin neurons were in the dorsal half of the nucleus, whereas phasically firing neurons were recorded throughout the nucleus, with more located ventrally (Fig. 2). This correlates well with recent immunohistochemical studies (Rhodes et al. 1981a).

In the PVN, albeit in a rather limited survey, we were unable to observe any marked differential distribution in the vertical plane of the oxytocin and phasic (vasopressin) neurons. A greater proportion of the neurons which were identified antidromically were found in the dorsal part of the PVN, however, which corresponds to the anatomical organization of this nucleus. Irrespective of where the oxytocin neurons were recorded they showed the typical high-frequency bursts of potentials prior to milk ejection as described by Wakerley and Lincoln (1973). During these bursts, which last 1–4 s, the frequency of firing is elevated some 20-

to 40-fold and the activity of individual neurons reaches 80 spikes/s in some cases. The peak activity is usually attained within 1 s, and the frequency often starts to decrease, as does the action potential amplitude, towards the end of the burst. Exactly how this stereotyped and synchronized activity is generated is still not clear. Is the burst caused by an intense depolarizing stimulus or by the removal of inhibitory damping from an otherwise paroxysmally excitable population of neurons? Is the synchrony due to a high degree of interneuronal connectivity or to the arrival of a synchronized afferent input? Recent anatomical data favour the view that some, at least, of the magnocellular neurons are closely linked. Connexions have been reported between the PVN and its contralateral paraventricular and ipsilateral SON (Silverman et al. 1981), and terminals containing neurosecretory material have been observed in the PVN and SON (Sofroniew and Glasmann 1981). Another interesting observation is that there is close apposition of some of the neuronal membranes within the PVN and these become more numerous during pregnancy and lactation (Theodosis et al. 1981). Electrophysiological evidence favours the concept that the neurosecretory burst is formulated remotely and arrives as a synchronized afferent barrage. The activity of individual oxytocin neurons prior to the neurosecretory burst does not appear to be synchronized to that of adjacent neurons, and though the size of the neurosecretory burst is related to the magnitude of the afferent stimulus, the magnitude of the response of individual neurons in no way relates to that of its neighbours. Furthermore, other froms of excitatory drive to these neurons, such as osmotic stimulation and chemical excitation (see Sect. 4.2 and 4.5), do not result in high-frequency bursts but rather in continuous activity, even though the frequency of discharge may be similar to that prior to the milk ejection response.

2.2.3 Vasopressin Neurons and the Genesis of Phasic Activity

Phasic activity has been routinely observed in recordings from magnocellular neurons; first reported by Wakerley and Lincoln (1971) for the rat, it was subsequently observed in the monkey (Hayward and Jennings 1973), the sheep (Jennings et al. 1978) and the rabbit (A. C. Paisley, personal communication). The activity typically consists of periods of regular discharge at 4–20 spikes/s interspersed with periods of silence. The duration of both the active and the quiescent phases is variable, both for individual neurons and between neurons (for a more detailed account see Poulain and Wakerley 1982). Unlike the oxytocin neurons there is no profound synchrony of activity between phasic neurons, though adjacent neurons will sometimes show a simultaneous onset of activity (Fig. 3). This type of activity is seen in its most dramatic form when the neurons are being driven intensely by stimuli known to release vasopressin, such as dehydration (Wakerley et al. 1978) or haemorrhage (Poulain et al. 1977); under these conditions the majority of the non-oxytocin neurons display phasic activity. Under normal conditions of hydration relatively few of these neurons display such activity and the distinctive on/off nature of the discharge is not so clearly marked; there are often a few sporadic potentials between bursts (Fig. 3). As discussed in Sect. 2.2.1 it is now generally accepted that the majority of neurons which display phasic activity are indeed vasopressin neurons. Certainly, stimuli which result in

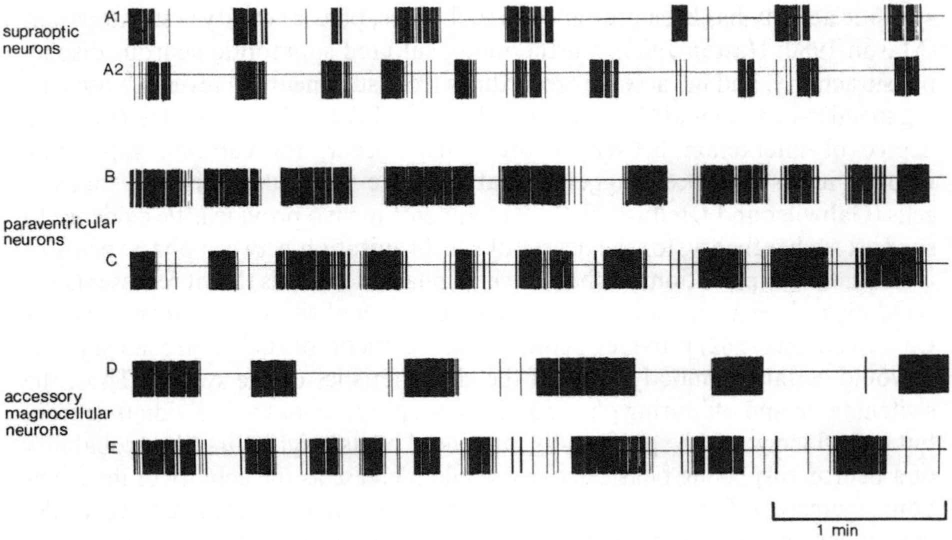

Fig. 3 a–e. Typical phasic activity recorded from six antidromically identified magnocellular neurons in different areas of the hypothalamus. **a 1** and **2** Two neurons firing phasically recorded simultaneously with a single microelectrode in the SON; note the small degree of synchronization. **b, c** The activity of two phasic neurons recorded in separate experiments in the PVN. **d, e** The activity of two neurons recorded in separate experiments from a small group of neurons located approximately 500 μm dorsal and 200 μm medial to the dorsal surface of the SON. Both these neurons and many of the adjacent neurons in this region could be identified antidromically when the neural lobe was stimulated

the secretion of vasopressin, such as osmotic challenge (Brimble and Dyball 1977), dehydration (Wakerley et al. 1978), haemorrhage (Poulain et al. 1977) or bilateral carotid occlusion (Dreifuss et al. 1976), profoundly increase the number of neurons displaying phasic activity and excite those neurons already phasically discharging. Unfortunately, these stimuli also excite oxytocin neurons, though the number identified that also display phasic activity is very small. The general pattern of evolution of phasic activity tends to be a progression from slow, irregular activity to a fast, continuous discharge which then breaks up into the classical phasic pattern. Neurons showing phasic activity during mild osmotic stimulation usually change to a continuous discharge following a further osmotic challenge; yet, curiously, vasopressin neurons recorded under conditions of severe chronic osmotic stress, e.g. during dehydration, invariably display phasic rather than continuous activity. The reason for this apparent anomaly may lie with the mechanisms responsible for phasic activity and may depend upon the proportion of neurons within the nucleus that are firing in this fashion.

There has been extensive debate as to the origin and evolution of phasic activity in the magnocellular neurons; it now seems likely that it is an intrinsic property of the neurons themselves. Phasic activity can be recorded from neurons in hypothalamic slices (Fig. 10) which are devoid of many of their afferent inputs; this suggests that major input from remote sites in not essential. Similarly, when synaptic activity has been prevented in such slices, phasic activity is still apparent (Mason 1980; Hatton 1982). Furthermore, cultured supraoptic neurons display phasic activity, and intracellular recordings from such neurons reveal an oscillating membrane potential (Gahwiler and Dreifuss 1979, 1980). Nevertheless, some degree of interaction between neurons may occur; for cultured supraoptic neurons it has even been suggested that there are pacemaker cells and follower cells (Gahwiler and Dreifuss 1979). Experiments in vivo provide little evidence to support such a theory, for the degree of synchronization between phasic neurons is small. One explanation for the genesis of phasic activity is that it represents the build up and decay of some material, possibly even of simple ions such as K^+ or Ca^{2+}. Hatton (1982) provides a convincing argument for such a mechanism, and it would certainly embody most of the characteristics of the system. Thus, the switching on and off during phasic activity is a property of the individual neuron, but one which would be profoundly influenced by its neighbours (the probability of a neuron displaying phasic activity would increase as the activity of its neighbours increases). This interaction could be further enhanced by increasing the degree of neuron-to-neuron apposition, by changes in the ultrastructure of glial elements, for example – a process reported to occur during dehydration (Tweedle and Hatton 1977). Such a system could still be triggered or modulated by synaptic influences and, depending on the ion channels influenced, different neurotransmitters might well evoke different forms of excitation, as has been reported for cultured neurons (Gahwiler and Dreifuss 1980).

2.2.4 Activity in Relation to Cortical EEG

Many hypothalamic neurons, like those in other brain areas, show changes in activity which relate to changes in cortical EEG; this in turn reflects the level of arousal of the animal. The most dramatic change in terms of behaviour of the animal and activity of the EEG is the transition from sleep to alert wakefulness. The cortical EEG changes from synchronous activity during sleep displayed as large-amplitude slow waves (<3 H_z), to desynchronized activity [high-frequency (>5 H_z), low-amplitude activity] during wakefulness. Conveniently, these patterns of EEG activity are maintained under light urethane anaesthesia which allows the concomitant neuronal activity to be recorded (Fig. 4). Lincoln (1969) found that approximately half of the neurons he recorded in the hypothalamus showed a change of activity with the transition of the EEG from slow-wave synchronized to desynchronized fast activity. Roughly a quarter of the total population showed greatest activity during the synchronized EEG phase, the other quarter firing faster during desynchronized EEG activity.

Our experience from recording vasopressin neurons is that they appear to be functionally specific, responding only to stimuli which release vasopressin, and do not reflect changes in the cortical EEG (Fig. 4); Wakerley and Lincoln (1971) reported similar observations. Intracarotid injections of hypertonic saline, which excite vasopressin neurons and release vasopressin, cause EEG arousal (desynchronized activity). However, we have noticed that intraventricular injection of hypertonic saline, a stimulus which also effectively excites vasopressin neurons (see Sect. 4.2), has no effect upon the cortical EEG. Thus, the arousal following

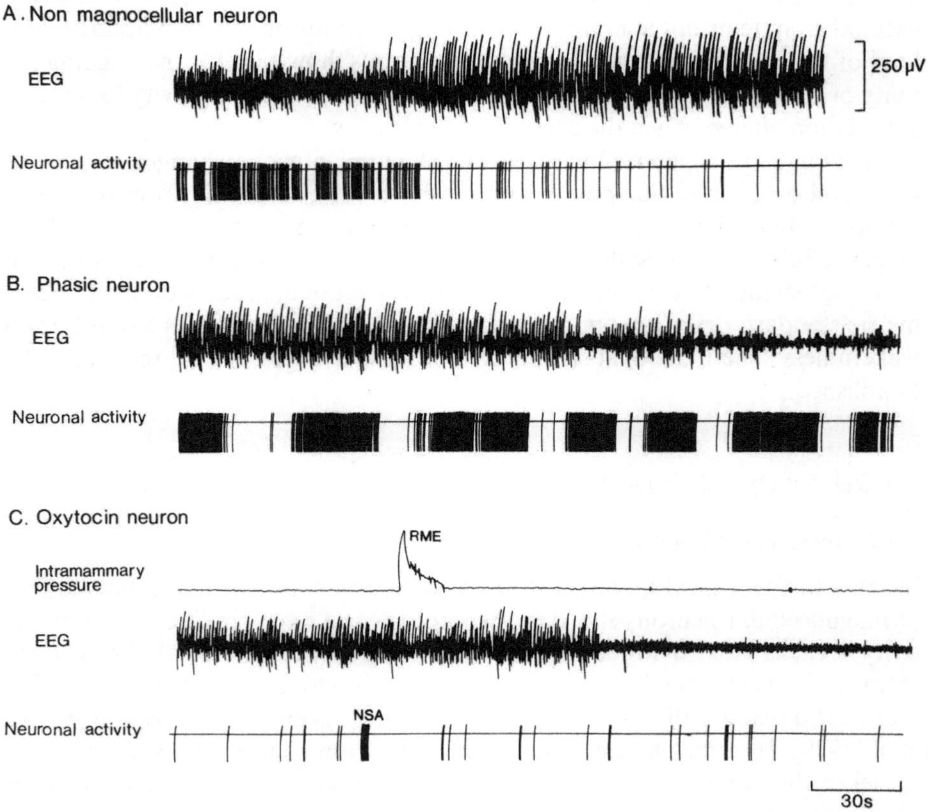

Fig. 4a–c. Simultaneous electrical recordings of cortical EEG and neuronal activity. **a** Activity of a hypothalamic neuron, recorded in the vicinity of the PVN, which could not be antidromically activated. Note the dramatic reduction in the rate of discharge of the neuron as the cortical EEG changed from desynchronized low-amplitude activity to large-amplitude, slow-wave activity. **b** A magnocellular neuron in the PVN firing phasically showed no alteration in activity as the EEG changed. **c** An oxytocin neuron recorded in the PVN which was firing very slowly showed no change in activity as the EEG altered. Note that the high-frequency burst of activity which preceded the rise in intramammary pressure occurred during a period of large-amplitude, synchronized slow-wave activity

intracarotid injections is a feature of the route of administration and not a necessary requirement for vasopressin release. The ongoing activity of oxytocin neurons also appears to be unrelated to the state of the cortical EEG; functional release of oxytocin, at least during suckling, does depend, however, upon the level of arousal of the animal. Thus, in the rat, activation of oxytocin neurons and release of the hormone will occur only when the EEG displays a slow-wave sleep pattern (see Fig. 4; Voloschin and Tramezzani 1979; Lincoln et al. 1980). An association between arousal and oxytocin release has also been observed in other species, but in the pig (Poulain et al. 1981 a) and the rabbit (Neve et al. 1982) oxytocin release during suckling occurs only during arousal. This reflex nicely demonstrates the capability and adaptability of the magnocellular neurons to interact with other systems in the brain. It seems likely that this interaction occurs at the level of the afferent input to the oxytocin neurons, however, for the ongoing activity of the neurons does not appear to change, nor is their sensitivity to osmotic stimulation altered, when the cortical EEG changes.

Although the magnocellular neurons show no obvious changes in activity with level of arousal, Summerlee and Paisley (1982), recording from conscious rats and rabbits, have reported subtle changes in the pattern of discharge of the magnocellular neurons without any overt changes in firing rate. There appear to be a lengthening of the modal interval and a corresponding reduction in the geometric standard deviation about this interval as the animal moves from sleep to wakefulness. The functional significance of these changes, if any, remains to be established.

2.3 Related Hypothalamic Neurons

2.3.1 Accessory Magnocellular Neurons

Immunohistochemical studies have revealed the existence of a substantial number of magnocellular neurons showing vasopressin- and oxytocin-like immunoreactivity in areas other than the SON and PVN (Rhodes et al. 1981 a). Axons from many of the neurons in these accessory nuclei project to the neurohypophysis, and represent possibly half of the hypothalamo-neurohypophyseal projection (Fisher et al. 1979). In principle it is feasible to record from neurons in these accessory nuclei in the same way as recording are made in the PVN and SON; the main problem is that these accessory nuclei are small and contain few magnocellular neurons. Inadvertently, however, we have recorded some neurons which were outside the confines of the SON. A total of 62 neurons that could be antidromically driven when the neural lobe was stimulated were recorded in an area dorsal to the SON. They were located as a distinct group, whose area of greatest density was some 500 μm dorsal to the SON and invariably some 200 μm medial to the main nucleus. Although not histologically identified, these would probably correspond to neurons in the vicinity of blood vessels which run from the SON to the PVN (see Ambach and Palkovits 1979). These neurons were similar in every way to those recorded in the SON, the threshold and latency for antidromic stimulation in the normal range for magnocellular neurons. Of the total neurons encountered, 43% showed phasic activity (see Fig. 3) and a further 17% were identified as oxytocin neurons by their response at milk ejection. Both the phasically active and the identified oxytocin neurons were excited by intraventricular sodium chloride and carbachol (see Sect. 4.2. and 4.5.1). It thus appears that these accessory neurons resemble the magnocellular neurons of the SON and PVN not only histologically but also electrophysiologically and functionally.

2.3.2 Atypical Magnocellular Neurons

The advent of immunohistochemical techniques has revealed the existence of anomalous neurons which do not conform to the conventional configuration of hypothalamic magnocellular neurons. Not all the magnocellular neurons stain for vasopressin and oxytocin (Vandesande and Dierickx 1975), and it has been suggested that some of them (perhaps those projecting to the neurointermediate lobe) may contain GABA (Vincent et al. 1982; Oertel et al. 1982) or opioid peptides (Rossier et al. 1979). If these neurons do indeed project from the magnocellular nuclei to the neural lobe they will be identified antidromically. This leads to speculation about their activity or function. Should we expect to find a third population of neurons during electrophysiological experiments, or do they have activity similar to that of the oxytocin or vasopressin neurons? We, like others (Brimble et al. 1978), have found a small population of cells which neither respond at milk ejection nor show phasic activity when driven by an osmotic stimulus. It is impossible to say at present whether these represent the anomalous neurons. Two basic assumptions would be required: first, that all oxytocin neurons respond prior to every milk ejection and, second, that all vasopressin neurons can be made to fire phasically. This has yet to be proved beyond doubt.

Another fascinating observation which has arisen from the application of modern histological techniques is the presence within the PVN and SON of neurons containing vasopressin or oxytocin which do not project to the neurointermediate lobe. Many are of the parvicellular type, but there are a few conventional magnocellular neurons in this group. Fibres found to terminate in the median eminence (Vandesande et al. 1977) are believed to be involved in regulating ACTH release. Recently, the newly identified corticotropin-releasing factor has been localized in the PVN, and sometimes may coexist with oxytocin (Swanson et al. 1983). Other vasopressin-, oxytocin-, or neurophysin-containing fibres have been found projecting to several forebrain areas, the brain stem and the spinal cord (Buijs et al. 1978; Sofroniew and Weindl 1978; Nilaver et al. 1980), and have been implicated in memory functions, emotions and behaviour, cardiovascular reflexes and regulation of the autonomic nervous system. Electrophysiological experiments are only just beginning to answer some of the questions raised by these histological observations. It appears that only a small number of these neurons also project to the neurohypophysis (Blume et al. 1978; Zerihun and Harris 1981) and, although a few neurons projecting to the median eminence showed phasic activity (Blume et al. 1978), the majority probably do not show the characteristics of the classical magnocellular neurohypophyseal neurons (Pittman et al. 1982).

3 Neurosecretion

One of the fundamental processes in the functioning of the magnocellular neurons is the mechanism of stimulus-secretion coupling, i.e. the translation of electrical activity into the release of precise quantities of neurosecretory product. Conveniently, this mechanism operates at the level of the nerve terminals and can thus be studied in detail by removing the neural lobe from the brain and treating it as an in vitro preparation (Douglas and Poisner 1964). Following depolarization of the terminals a series of biochemical and morphological changes are triggered which result in the release of the hormones. In brief, the depolarization causes a brief influx of calcium ions which eventually leads to the fusion of neurosecretory granules and the cell membrane, with the consequent release of granule content by exocytosis (see Douglas 1974; Morris et al. 1981). The granule membrane is believed to be taken back within the terminal (Theodosis 1982), and increased intracellular calcium is sequestered into microvesicles which may themselves be subsequently released (Shaw and Morris 1980; Nordmann and Chevallier 1980). It is generally assumed that action potentials arriving at the terminals are responsible for the depolarization. Certainly, the release of hormones can be mimicked by electrical stimulation of the pituitary stalk, both in vivo and vitro, and release evoked by stimulation of the pituitary stalk is prevented by the blocking of action potential propagation (Dreifuss et al. 1971; Nordmann and Dreifuss 1972). Further, as indicated in Sect. 3.1 the activity of the neurons and the profile of release are well correlated. The arrival of action potentials may not be the only mechanism responsible for secretion, however, for the terminals can be depolarized in vitro by potassium ions alone in the absence of action potentials (Douglas and Poisner 1964; Dreifuss et al. 1971). Thus, locally released or blood-borne substances might also trigger neurosecretion. It has recently become apparent that a considerable degree of influence may be exerted by endogenous peptides and monoamines acting on the terminals to modulate the release of neurosecretory product, either in conjunction with or in the absence of action potentials (see Sect. 3.2).

3.1 Significance of the Firing Pattern for Neurosecretion

One of the most distinctive features of the magnocellular neurons is their characteristic firing patterns at the time that they are actively secreting, namely the high-frequency bursts of the oxytocin neurons during suckling and the phasic activity of the vasopressin neurons when blood osmolality is raised. Detailed study of these activity patterns has revealed that it is not only the frequency of discharge which is important for release but also the temporal organization. By electrically stimulating the neural lobe either in vitro or in vivo it can be demonstrated that hormone release increases with the frequency of stimulation (Harris et al. 1969). This is not due merely to an increase in the number of depolarizations in unit time, for the facilitation of release still occurs if the stimulus duration is reduced, when the frequency is increased such that the total number of stimuli remains constant (see Fig. 5 Dreifuss et al. 1971). A more detailed examination of the quantity of oxytocin released by electrical stimulation in vivo demonstrated that the most effective stimuli were in the range of 2–4 s at 40–60 Hz (Lincoln et al. 1976; Dreifuss et al. 1981), precisely the range observed for the neurosecretory bursts of oxytocin neurons during suckling (Wakerley and Lincoln 1973). It has been calculated that at these frequencies the ability of the individual potentials to release hormone is enhanced some 100-fold compared with a discharge rate of 1 spike/s (Lincoln 1974).

Dutton and Dyball (1979) have suggested that there is potentiation of release with increased frequencies of stimulation at vasopressin nerve terminals, though the optimal frequency is lower at about 6 pulses/s. The same authors found that

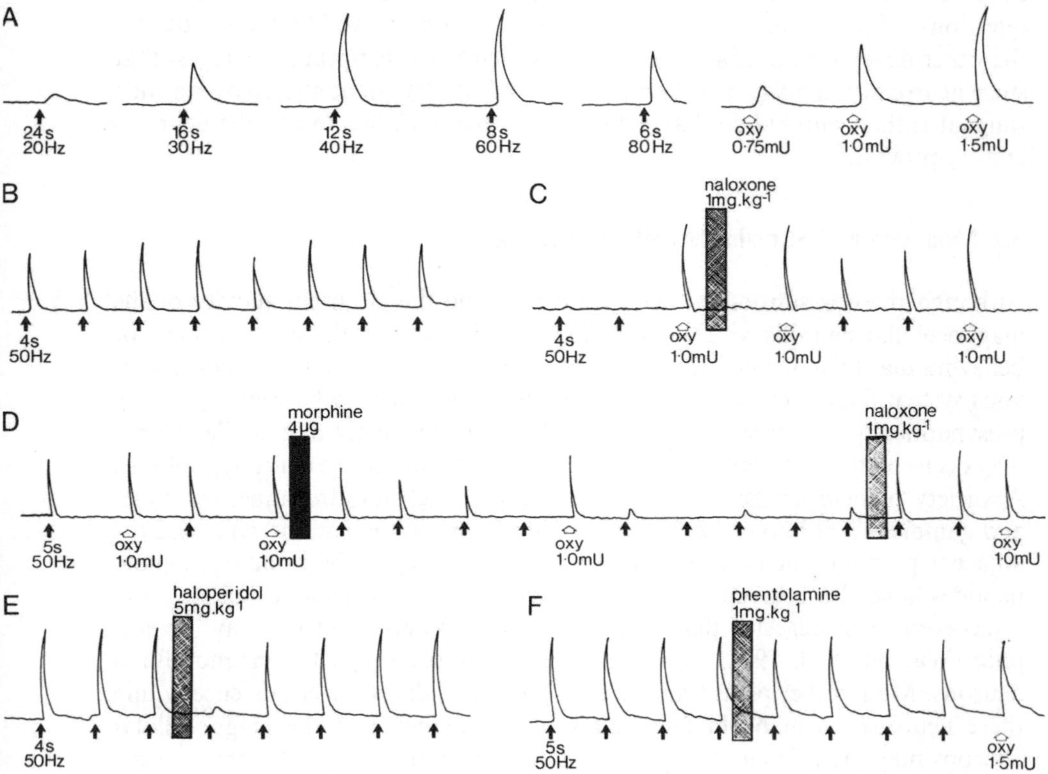

Fig. 5a–f. Intramammary pressure responses in the anaesthetized rat, produced by pulses of oxytocin released from the neural lobe when it was electrically stimulated every 3 min (*black arrows*). **a** Shows that stimulation between 40 and 60 Hz is optimal for oxytocin release. The stimuli, all consisting of 480 pulses, were applied in a random sequence but have been arranged in order for clarity. Oxytocin was injected intravenously (*open arrows*) to determine mammary sensitivity. **b** Shows the reproducible nature of neural lobe stimulation. **c** The rat initially failed to release oxytocin when electrically stimulated, but following the intravenous injection of naloxone (1 mg/kg^{-1}) normal release occured. **d** Intraventricular morphine (4 μg) suppressed, and intravenous naloxone restored, the release of oxytocin evoked by electrical stimulation of the neural lobe. There was no change in mammary sensitivity to oxytocin. **e** Immediately after the injection of the dopamine antagonist haloperidol (5 mg/kg^{-1}) the electrically evoked release was blocked, but it returnd 3 min later. **f** The adrenocepter antagonist phentolamine had no effect upon electrically evoked release

stimulating the isolated neural lobe with a pattern of stimuli which mimicked the activity of a phasically firing neuron was even more effective than the same number of pulses at regular intervals. They attributed this facilitation to the preponderance of short interspike intervals in the phasic stimulus. It has also been noted that this enhancement of vasopressin output is not constant; the system becomes fatigued and less efficient after about 2 min (Ingram et al. 1982). The majority of neurons discharging phasically either in vivo or in vitro have active periods of less than 2 min, which again reveals the ideal relationship between endogenous electrical activity and optimal parameters for secretion. The fascinating questions, of course, are: Which physiological process evolved first, and how did the other develop to produce the optimal system? It is tempting to suggest that since neuronal activity (generation of action potentials) can be altered over a wide range it is this neuronal discharge which has been modelled to suit the neurosecretory process.

3.2 Modulation of Stimulus-Secretion Coupling

Although there is substantial evidence correlating the electrical activity of the magnocellular neurons with the secretion of their product, there are reasons for believing that this may not always be the case. In many areas of the central nervous system there is evidence of inhibitory or facilitatory mechanisms operating presynaptically on nerve terminals, and it now seems likely that similar events may occur on the terminals of the magnocellular neurons in the neurohypophysis. A variety of neurotransmitters, including dopamine, noradrenaline, serotonin and gamma aminobutyric acid (GABA), are found in the pars nervosa and the adjacent pars intermedia (the neurointermediate lobe) of the pituitary. Opioid peptides have also been demonstrated by immunohistochemical techniques, and it has even been suggested that enkephalins (Martin and Voigt 1981) and dynorphin (Watson et al. 1982) may coexist with vasopressin in the magnocellular neurons. Many other peptides have also been candidates for coexistence within these neurons, so many in fact that some components of the magnocellular neurons may provide an attractive but nonspecific binding site for the antisera used in immunohistochemical investigations (van Leeuwen 1980).

3.2.1 Monoamines

Of the catecholamines, dopamine is predominant in the neural lobe (Saavedra et al. 1975); dopamine receptors are present (Stefanini et al. 1980) and a dopamine receptor-linked adenylate cyclase has been reported (Ahn et al. 1979). Several groups have investigated the effect of dopamine on vasopressin and oxytocin release. Bridges et al. (1976) reported that dopamine stimulated the release of both vasopressin and oxytocin from the isolated lobe but that very high doses were required. Others have reported a suppression of the secretion of vasopressin (Lightman et al. 1982) and oxytocin (Vize and Volbekas 1980). An explanation for this anomaly has been provided by Racké and colleagues (1982a). They observed both a facilitation and a suppression of vasopressin release by dopamine; the facilitation was mediated through D1 receptors blocked by *cis*-flupenthixol, while the inhibitory action was mediated by D2 receptors and blocked by sulpiride. The hypothesis that release is facilitated by D1 receptors (an adenylate cyclase-linked system) fits with the observation that at times of vasopressin release the levels of cyclic nucleotides in the neural lobe are elevated (Vale and Hope 1982). [Our own experience that haloperidol (a D1 antagonist) caused a short-lived suppression of the release of oxytocin evoked by electrical stimulation of the neural lobe in vivo (Fig. 5) would fit the concept of D1 receptor-linked facilitation of release, though a transient change in blood flow through the lobe might be the more likely explanation.] Whether the effects of dopamine on the isolated neural lobe reflect endogenous physiological mechanisms is not clear. The levels of dopamine in the neural lobe do show diurnal variation, however (Barden et al. 1982), and are influenced by lactation and dehydration (Holzbauer et al. 1978; Alper and More 1982).

Noradrenaline is also present in the neural lobe and can be released along with dopamine by depolarization (Sharman et al. 1982); it is mainly but not completely accounted for by sympathetic fibres from the superior cervical ganglion, stimulation of which has been reported to stimulate the release of neurohypophyseal hormones (Karasek et al. 1980). Similarly, the β-adrenoceptor agonist isoprenaline has been reported to stimulate the release of vasopressin via an adenylate cyclase system which was distinct from the dopamine-linked system mentioned previously (Racké et al. 1982b). In our opinion, it seems unlikely that a similar system would facilitate the release of oxytocin, for propranolol (a β-adrenoceptor antagonist) facilitates the release of oxytocin during milk ejection (Tribollet et al. 1978). Rather, we have suggested that an adrenergic system acting on β-adrenoceptors at the level of the lobe might be inhibitory to the release of oxytocin (Tribollet et al. 1978). Recent experiments (unpublished data) in which we have applied isoprenaline to the electrically stimulated neural lobe in vitro do not support such a hypothesis, however.

3.2.2 Opioid Peptides

The most convincing evidence to date that putative neurotransmitters can influence the release of neurohypophyseal hormones at the level of the terminals is that of opioid-mediated inhibition of oxytocin release in the rat. We first became aware of this phenomenon in experiments in which oxytocin neurons were being recorded during suckling as small doses of morphine were injected intraventricularly (Clarke et al. 1979b). Although the high-frequency neurosecretory bursts continued and were unaffected, the mammary-pressure responses (and thus milk ejection) ceased (see Fig. 11). As the mammary sensitivity to oxytocin was unaltered, the only explanation was that secretion had been prevented. We could also demonstrate that the release of hormone evoked by electrical stimulation of the neural lobe in vivo was blocked by opioid peptides and morphine (Fig. 5), again suggesting that the site of action was the nerve terminals. There is anatomical evidence to support this suggestion, for enkephalin-containing fibres are present in the neurointermediate lobe (Rossier et al. 1979), as are abundant opiate binding sites (Simantov and Snyder 1977). We recently obtained unequivocal evidence by demonstrating that morphine would suppress the release of oxytocin evoked electrically from the isolated neural lobe in vitro (see Fig. 6; Clarke and Patrick 1983). Endogenous opioid peptides can initiate this inhibitory influence, for naloxone will potentiate oxytocin release both in vitro (Bicknell and Leng 1982; Clarke and Patrick 1983) and in vivo (Fig. 5; Clarke et al. 1979b). We have also noticed that the inhibitory effect of the opiates is more pronounced when a high-frequency stimulus is used to evoke release. This is not surprising, perhaps, since frequency facilitation is believed to result from improved calcium conductance and the opiates are thought to interfere with calcium transport (Chapman and Way 1980).

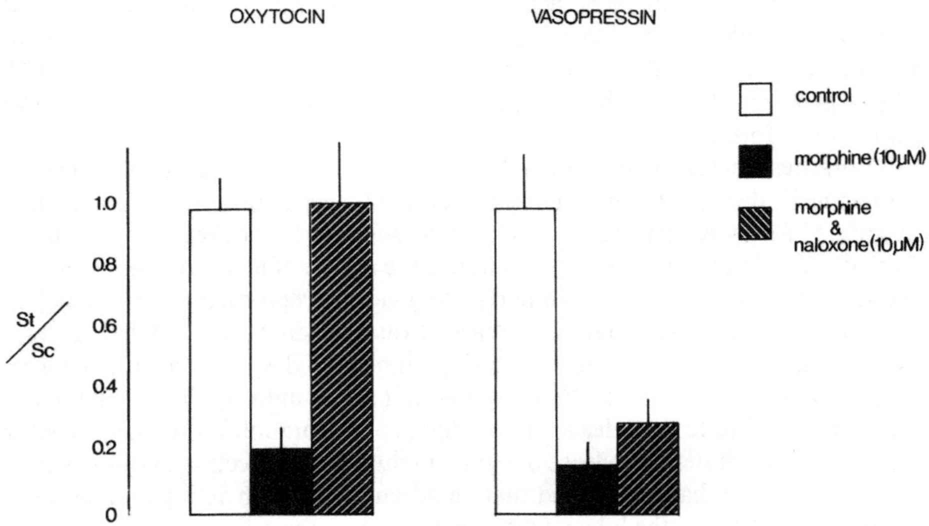

Fig. 6. Effect of morphine (10 μ*M*) on electrically evoked release of vasopressin and oxytocin from the rat neurointermediate lobe in vitro. Morphine suppressed the release of both vasopressin and oxytocin, but only the inhibition of oxytocin release was reversed by naloxone. The lobe was dissected and prepared according to the method of Douglas and Poisner (1964). Every 20 min the incubation medium (0.5-ml sample) was changed and prior to immersion in alternate samples the lobe was stimulated electrically through the pituitary stalk (500-μA, 1-ms pulses at 30 Hz for seven periods of 10 s, every 20 s). Morphine was added to the incubation medium 20 min before the second period of stimulation, and morphine and naloxone 20 min before the third period of stimulation. Results are expressed as *St*/*Sc* which is the ratio of the quantity of vasopressin or oxytocin released (measured by radioimmunoassay) in the presence of the test substance to that released in the control medium

The evidence for opioid peptides regulating vasopressin release at the level of the neural lobe is not so clear. Unfortunately, there are no suitable stimuli which can be guaranteed to release solely vasopressin, and detection of the released hormone in vivo during an electrophysiological experiment is not as simple as detection of oxytocin. Thus, the only evidence to date is that obtained using the isolated neural lobe preparation. Using this technique data showing that opioid peptides will inhibit the release of vasopressin have been obtained (Iversen et al. 1980; Lightman et al. 1982). However, other groups have obtained conflicting results (Christensen and Fjalland 1982; Bicknell and Leng 1982). Our own results suggest that if the release of oxytocin and that of vasopressin are measured simultaneously from the same lobe, morphine will inhibit the release of both peptides to the same extent, but only the suppression of oxytocin secretion can be reversed by naloxone (Fig. 6). Whether this observation reflects the involvement of different types of opiate receptors which have different naloxone sensitivity, or of an intermediary substance (such as dopamine) is not clear.

4 Factors Affecting the Activation of Magnocellular Neurons

4.1 Suckling and Related Stimuli

As described in Sect. 2.2.1 the stimulus provided by the sucking of the young on the nipples to release pulses of oxytocin provides, at least in the rat, the most specific stimulus yet studied for the activation of a subgroup of the magnocellular neurons. The characterization of the reflex activation of oxytocin neurons has been extensively researched since the pioneering studies of Wakerley and Lincoln (1973). The essential characteristic is the integration of the continuous stimulus provided by the sucking of the young, at one or more unknown sites within the brain, such that pulses of oxytocin (equivalent to 0.5–1.0 mU of oxytocin injected intravenously) are released every 5 min or so for as long as the young remain on the nipples. At the level of the oxytocin neurons this reflex activation is expressed as a short-duration (1–4 s), high-frequency (10–80 spikes/s) burst of potentials which consistently precedes the mammary-pressure response by some 12–20 s. The size of the neurosecretory burst is dependent upon the afferent input, namely the number of young sucking on the nipples (Lincoln and Wakerley 1975). The magnitude of the response of an individual neuron does not necessarily relate to that of its neighbours but does appear to reflect its own ongoing activity; thus, the higher the background activity, the larger the neurosecretory response. This ongoing activity of oxytocin neurons, however, is not influenced by suckling in the anaesthetized rat. Similar neurosecretory bursts have been recorded in the unanaesthetized rat during suckling (Summerlee and Lincoln 1981), and recordings in the conscious rabbit have revealed short, high-frequency bursts superimposed on a low background activity during suckling (A. C. Paisley, personal communication). Although many attempts have been made to delineate the afferent pathway for suckling-evoked oxytocin release (for details see Poulain and Wakerley 1982), little is known about the central site at which the timing and magnitude of the neurosecretory activation of oxytocin neurons are determined.

Oxytocin is also believed to be released during labour in response to the stretching of the cervix (Ferguson reflex). Experiments involving artificial dilation of the cervix have proved rather disappointing; oxytocin neurons are excited briefly, but so are the phasic neurons (Dreifuss et al. 1976), and the intensity of the stimulus required is such that a nonspecific stress response cannot be excluded. Anaesthesia causes termination of labour in the rat, but it has been possible to record from identified magnocellular neurons during labour in conscious rats chronically implanted with electrodes (Summerlee 1981). In these studies a proportion of the identified magnocellular neurons gave high-frequency bursts of activity during labour, which were correlated with abdominal straining; some of these neurons were recorded through to the nursing period and gave bursts of activity prior to reflex milk ejection. There was also an increase in phasic activity during labour.

4.2 Osmotic Stimulation

The main physiological role of vasopressin is that of an antidiuretic on the kidney, and as would be expected an increase in osmolality of the blood provides a powerful stimulus for the release of vasopressin. A variety of approaches have been used to investigate this action. Chronic methods have involved dehydration by withholding water, or osmotic loading by providing 2% saline to drink. Acute stimulation has been achieved by injecting hyperosmotic solutions of sodium chloride or other solutes either intraperitoneally or directly into the blood stream by intravenous or intracarotid injection. Each of these techniques results in an increased level of vasopressin in the plasma and an increase in the activity of the putative vasopressin neurons (for a detailed survey see Poulain and Wakerley 1982). Essentially, as the osmotic pressure of the blood increases so does the activity of the magnocellular neurons, and the proportion of neurons showing phasic discharge also increases dramatically (Brimble and Dyball 1977; Wakerley et al. 1978). The activity of the magnocellular neurons is reduced following a water load. One complicating aspect of studying the effect of osmotic stimulation on the activity of the magnocellular neurons is that these stimuli also effectively excite the oxytocin neurons (Brimble and Dyball 1977; Wakerley et al. 1978) and release substantial quantities of oxytocin into the circulation (Dyball 1971; Brimble et al. 1978). Whether this release is a physiological process [oxytocin has been shown to have a natriuretic effect on the kidney when the sodium ion concentration in plasma is high (Balment et al. 1980)] or merely reflects the common ancestry of vasopressin and oxytocin neurons remains obscure.

Many of the characteristics of osmotic regulation were elegantly demonstrated by Verney in 1947, and in some respects little advance has been made since. Two major areas of controversy still exist: (a) Where is the osmoreceptor situated! and (b) Do the osmoreceptors respond to osmolality in general or specifically to the concentration of sodium ions in the blood and/or cerebrospinal fluid? The diversity of opinion, based mostly on acceptable experimental evidence, suggests that there are probably multiple sites which are osmosensitive. The vasopressin neurons themselves appear to be responsive to changes in sodium ion concentration. In experiments in vivo in which sodium ions were applied directly onto cell membranes all magnocellular neurons were excited, while most of the adjacent hypothalamic neurons were not affected (Leng 1980). Similarly, in experiments in which the activity of supraoptic neurons was determined in the isolated hypothalamic slice in vitro an increase in neuronal activity was recorded when the osmotic pressure of the perfusing medium was raised (Mason 1980; Wakerley and Noble 1982). Raising the concentration of sodium ions was found to be more effective in this system than raising the osmotic pressure using other solutes, such as mannitol. Although raising osmotic pressure caused membrane depolarization in the slice preparation, it was rarely sufficient to initiate action-potential generation when synaptic transmission was blocked (Mason 1980). This suggests that although the magnocellular neurons may be osmosensitive, they also require a synaptic input for full expression of the osmotically evoked response.

Similar results have been obtained in vivo in rats in which the hypothalamus has been surgically isolated (Dyball and Prilusky 1981); in these animals the supraoptic neurons showed a low level of spontaneous activity and no response to an osmotic stimulus. The exact site of these other osmosensitive elements which provide the afferent input to the magnocellular neurons has been the subject of much debate, and extensive discussions have centred on whether these areas are on the vascular or the neural side of the blood-brain barrier (see Andersson 1978; Bie 1980). Certainly, intra-arterial, intravenous, and even intraperitoneal injections of hypertonic sodium chloride result in the excitation of vasopressin and oxytocin neurons. The circumventricular organs provide an attractive site for the location of osmosensitive structures, for these are highly vascular areas which – because of their ultrastructural organization – lack a conventional blood-brain barrier (Dellman and Simpson 1979). Of particular interest are the subfornical organ (SFO), the organum vasculosum of the lamina terminalis (OVLT), and parts of the wall of the anterior third of the third ventricle (AV3V). These structures receive and send projections to the SON and PVN (Weindl and Sofroniew 1981; Miselis 1981), and are known to be involved in thirst mechanisms and the release of vasopressin evoked by angiotensin (Simpson 1981).

In order to further study the effect of osmotic stimulation on the activity of magnocellular neurons we have made use of the technique of injecting hypertonic solutions into the ventricular system. This had previously been used on large species such as the goat and the dog, but we modified it to suit the rat (Clarke and Merrick 1982). Injection of 1–2 μl sodium chloride (1–3 *M*) into either the third ventricle or a lateral cerebral ventricle results in rapid and profound excitation of magnocellular neurons (Fig. 7). Over 90% of the identified magnocellular neurons so far encountered have been excited within 30 s by the stimulus and have shown a 2.5- to 10.5-fold increase in activity lasting 2–12 min. Of the adjacent, nonmagnocellular hypothalamic neurons studied less than 40% were excited by the stimulus, and the latency to onset was always greater than 60 s; the remaining hypothalamic neurons were either unaffected or actually inhibited. The stimulus also causes release of sufficient oxytocin and vasopressin to be detected by changes in intramammary pressure and blood pressure respectively. This effect of intraventricular sodium chloride is prevented by radio-frequency damage of the SFO.

This method of stimulation is a very convenient technique for studying osmotic activation of the magnocellular neurons. The stimulus is potent but only short lived, which allows it to be repeated several times if required; reproducible responses are obtained. This technique has enabled us to identify some of the neurotransmitters in the afferent pathway which drive the magnocellular neurons during osmotic stimulation (see Sect. 4.5). As the stimulus allows both vasopressin and oxytocin neurons to be driven through a range of firing frequencies it is possible to determine whether a neuron is capable of displaying phasic activity, even though it may be slowly firing or quiescent when first encountered. Some of the characteristics of the magnocellular neurons have been elaborated further. For instance, though oxytocin neurons show characteristic short-duration bursts during suckling, when stimulated osmotically they show a fast, continuous activ-

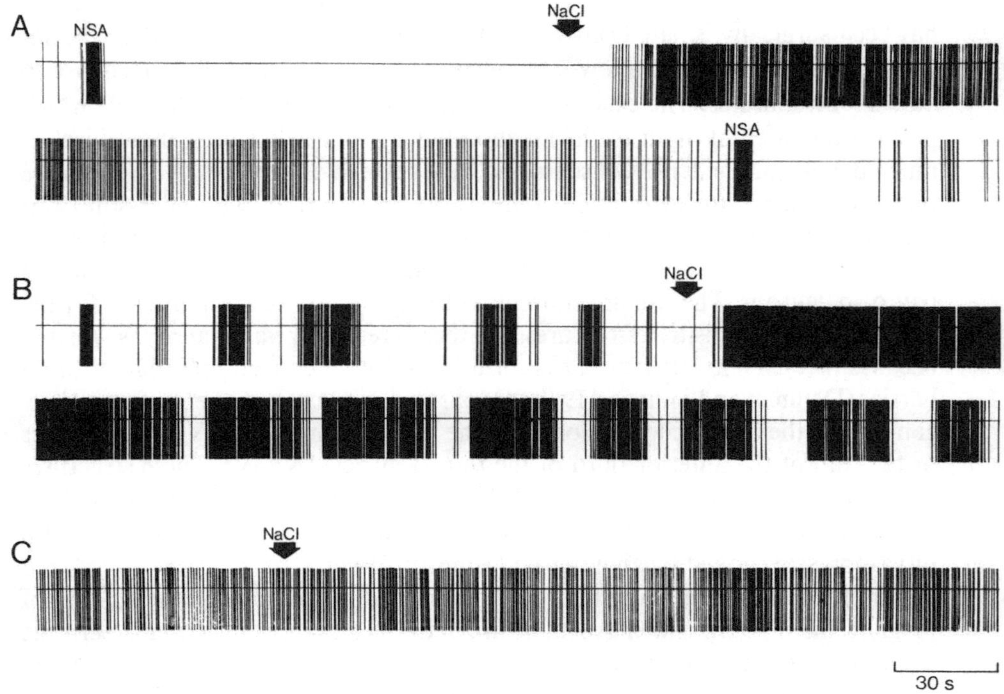

Fig. 7 a–c. The effect of hypertonic sodium chloride (1 μl, 3*M*) injected intraventricularly on the firing rate of three hypothalamic neurons. **a** An oxytocin neuron showing neurosecretory bursts of activity (*NSA*) during suckling is excited by hypertonic sodium chloride (*NaCl*). **b** A phasic neuron showing excitation (a change from phasic to continuous discharge) following hypertonic sodium chloride. **c** A hypothalamic neuron in the vicinity of the SON which could not be antidromically stimulated was not affected by sodium chloride. (In **a** and **b** the two rows are part of a continuous record)

ity which is sustained for several minutes even though the frequency of discharge may be as high as in the suckling-evoked neurosecretory burst. Although many of these neurons were profoundly excited by the intraventricular hypertonic stimulus, none of the oxytocin neurons changed activity to display a phasic pattern of discharge. This reflects the change occurring during severe dehydration, for although the majority of non-oxytocin neurons fire phasically, none of the those identified showed phasic activity after 18 h of dehydration, although they were significantly excited (Wakerley et al. 1978).

Based on evidence from our own experiments and from our colleagues, our understanding of the osmoregulatory system in the rat is somewhat similar at present to that expressed by Leng and his colleagues (1982). The supraoptic neurons are basically sodium sensitive, though probably not osmosensitive, and in the intact animal they are powerfully excited by afferent input from remote sites when exposed to an osmotic challenge. One such site is the SFO, which appears to be the primary site for detecting increased osmotic pressure in the CSF; again, sodium ions appear to be the most potent stimulus. Oxytocin neurons are as profoundly excited as vasopressin neurons, and the release of oxytocin may actually exceed that of vasopressin – the significance of the release of oxytocin remains obscure.

4.3 Renin-Angiotensin System

The renin-angiotensin system plays an important part in the regulation of water balance. Hypovolaemia results in renal renin release and in an elevation in the level of angiotensin II in the plasma, which in turn stimulates the thirst mechanism (see Andersson 1978). Under some circumstances the elevated plasma levels may be sufficient to stimulate the release of vasopressin (Share 1979), though it has been argued that physiological levels are insufficient to produce such an action (Reid et al. 1982). A further possibility is that the angiotensin, even at low concentrations, facilitates the sodium ion-evoked release of vasopressin (Buggy et al. 1979; Stamler and Phillips 1981). When injected intraventricularly, angiotensin certainly has a potent effect upon the release of vasopressin (Keil et al. 1975; Hoffman et al. 1977), which suggests that it may have a central site of action. The angiotensin may even be central in origin, for the existence of a separate brain renin-angiotensin system has been discussed (Reid 1979; Meyer et al. 1982; Ganten et al. 1984; Lind et al. 1985). A potential site of action for angiotensin, whether originating in the blood or central in origin, is the circumventricular organs. Of particular interest is the SFO, which is an important site for angiotensin-mediated drinking behaviour (Simpson and Routtenberg 1975), and shows macromolecular changes after dehydration or local application of renin (Summy-Long and Severs 1979). Angiotensin injected intraventricularly has been reported to excite both vasopressin and oxytocin neurons in the PVN and SON (Akaishi et al. 1980). A synergistic action between angiotensin and hypertonic sodium chloride was observed and the actions of angiotensin were blocked by the antagonist saralasin. Thus, angiotensin may excite the magnocellular neurons indirectly via periventricular structures. A more direct action has also been claimed for angiotensin, as it excites supraoptic neurons in vivo when applied iontophoretically (Nicoll and Barker 1971), and cultured supraoptic neurons when superfused (Gahwiler and Dreifuss 1980). angiotensin does not always excite neurons; in one study on the septum all those which responded to iontophoretically applied angiotensin were inhibited (Simonnet et al. 1980). Septal neurons have been shown to project to the SON (Poulain et al. 1981 b), and activation of them results in inhibition of firing of both vasopressin and oxytocin neurons (Poulain et al. 1980). Thus, there is also a potential mechanism whereby angiotensin may excite supraoptic magnocellular neurons by a process of disinhibition.

4.4 Haemodynamic Changes

For many years vasopressin was not thought to be a contributory factor in the regulation of vascular tone; however, it has recently been considered to have a significant influence on the control of blood pressure (Cowley 1982). Certainly, changes in cardiovascular pressure, by stimulation of the cardiopulmonary receptors (Gauer et al. 1970) and arterial baroreceptors (Share and Levy 1966), can have a profound effect upon the release of vasopressin. Two experimental techniques involving the vascular system have been used extensively to excite magnocellular neurons in electrophysiological experiments; these are haemorrhage (Poulain et al. 1977) and carotid occlusion (Harris et al. 1975; Dreifuss et al. 1976). The latter, claimed to be fairly specific for activating vasopressin neurons, results in a powerful but short-lived excitatory response, and in synchronization of the discharge of the vasopressin neurons. Haemorrhage, which may be achieved by passing a cannula into the right atrium of the heart and withdrawing up to 5 ml of blood, results in a slow but profound excitation of both vasopressin and oxytocin neurons. The rate of discharge remains high until the blood is returned, at which time neuronal activity returns to the previous level (Poulain et al. 1977). However, we preferred to have 1–3 ml of blood pumped out passively from a cannulated artery; using this procedure we observed a change in activity after 0.5–1 ml blood loss (Fig. 8). Over 70% of the magnocellular neurons identified (both vasopressin and oxytocin neurons) were excited by the removal of blood, yet less than 20% of nonmagnocellular hypothalamic neurons were affected. Replacement of the blood had a strong inhibitory effect and reduced the activity to its previous level, as reported by Poulain et al. (1977). Increasing the blood volume has a strong inhibitory action on the magnocellular neurons by stimulating the atrial stretch receptors (Koizumi and Yamashita 1978; Menninger 1979). Stimulation of the baroreceptors also results in an inhibition of neurons in the SON, whereas stimulation of the carotid chemoreceptors increases the activity of phasically firing neurons (Harris 1979). Information from these peripheral vascular receptors, like that from other visceral afferent fibres, is relayed by the solitary nucleus in the medulla. On the basis of anatomical and electrophysiological experiments a monosynaptic pathway has been proposed, projecting from the solitary nucleus to the PVN (Ciriello and Calaresu 1980). Reciprocal projections from the PVN to the medulla containing oxytocin and vasopressin exist as well. A few of these neurons in the PVN have been shown – by both anatomical (Swanson and Kuypers 1980) and electrophysiological techniques (Zerihun and Harris 1981) – to project to both the neurohypophysis and the medulla, and these could provide the basis for feedback control.

4.5 Neurotransmitters and Neuromodulators

Countless substances have been shown to influence the activity of the magnocellular neurons, or have been implicated in their regulation. Amino acids such as glutamate and GABA (Nicoll and Barker 1971; Moss et al. 1972b; Arnauld et al. 1983), serotonin (Moss et al. 1972b; Moos and Richard 1983), substance P (Clarke et al. 1980; Unger et al. 1981), ACTH (Sawchenko et al. 1982), estrogen (Rhodes et al. 1981b), histamine (Bennett and Pert 1974; Haas et al. 1975), and even prostaglandins (Hoffman and Schmid 1979; Hoffman et al. 1982) have been suggested. The amount of information regarding the involvement of these substances is limited, however, and the subject does not warrant detailed discussion

Fig. 8 a, b. The effect of blood removal and replacement on the acitivity of two antidromically identified supraoptic neurons. Trace **a** illustrates the effect of blood removal; during withdrawal the neuron fired faster, and the increased activity continued after the removal of the blood. Trace **b** illustrates the effect of replacing blood on the activity of a phasic neuron; as the blood was replaced the length of the quiescent phases increased and there was an overall reduction in activity

at present. With regard to four major classes of transmitters – acetylcholine, opioid peptides, catecholamines and the neurohypophyseal hormones themselves – there is substantial evidence for action on and involvement with the magnocellular neurons, and these will be considered in detail.

4.5.1 Acetylcholine

It has been thought for many years that acetylcholine might be involved in the release of vasopressin (Pickford 1939). Acetylcholine injected into the vicinity of the SON was found to cause antidiuresis (Pickford 1947), as was nicotine (Milton and Paterson 1974). Using organ-cultured hypothalamo-neurohypophyseal systems, Sladek and colleagues showed that acetylcholine and nicotine released vasopressin, and that this effect was blocked by "nicotinic receptor" antagonists (Sladek and Knigge 1977; Sladek and Joynt 1979). The application of drugs by microiontophoresis directly onto magnocellular neurons in vivo has demonstrated that these neurons could be excited by cholinergic nicotinic receptors (Dreifuss and Kelly 1972 a); inhibitory muscarinic cholinergic responses were also reported (Barker et al. 1971). In subsequent investigations in which the neurons were subdivided according to whether they showed phasic activity, it was observed that it was predominantly the phasically firing neurons which were excited by acetylcholine and nicotine (Bioulac et al. 1978; Arnauld et al. 1983). The predominant feature in each of these studies was the change from phasic activity to continuous discharge during, and for some time after, the acetylcholine application, but the overall firing rate was not increased. In contrast, glutamate initiated bursts of activity and increased the firing rate; ongoing phasic activity could still be observed during continuous glutamate applications. This suggests that the two transmitters may influence different ion channels, which may themselves be important in the genesis of phasic activity. If a cholinergic-nicotinic component is important for the transition of phasic to continuous activity, it follows that physiological stimuli which release vasopressin might be suppressed by nicotinic antagonists.

The release of vasopressin evoked by a hypotensive stimulus can be blocked by nicotinic but not by muscarinic antagonists (Bisset and Chowdrey 1981). We examined the effect of haemorrhage on magnocellular neurons and found that the nicotinic antagonist mecamylamine (4 mg/kg^{-1} i.v.) prevented the expected excitation in 86% of the neurons tested, whereas the muscarinic antagonist atropine (1 mg/kg^{-1}) had no effect. Whether the nicotinic synapse involved is at the level of the supraoptic neurons remains to be established; at least one other nicotinic site capable of releasing vasopressin has been localized on the ventral surface of the medulla (Bisset et al. 1975).

Osmotically stimulated release of vasopressin has also been prevented by nicotinic antagonists, both in vivo (Bridges and Thorn 1970) and in the cultured hypothalamo-neurohypophyseal preparation (Sladek and Joynt 1979), though others have failed to block release (Milton and Paterson 1974). We have prevented the excitation of vasopressin and oxytocin neurons evoked by intraventricular sodium chloride, and thus blocked the release of the neurohypophyseal hormones, but only by using a large dose of mecamylamine 15 mg/kg^{-1} intravenously, or by injecting the mecamylamine (50 μg) intraventricularly.

Cholinomimetics such as carbachol, injected into the cerebral ventricles, also provide a powerful stimulus for the release of vasopressin and oxytocin (Kuhn and McCann 1971; Kuhn 1974), but in this case the effect is blocked by muscarinic antagonists such as atropine. As the only muscarinic response is an inhibition of the putative oxytocin neurons (Barker et al. 1971; Arnauld et al. 1983), the intraventricular carbachol must exert its effect at a site remote from the magnocellular neurons.

We have undertaken a series of experiments to further elaborate the nature of the cholinergic involvement in the control of vasopressin and oxytocin neurons. Carbachol (0.2 μg), injected intraventricularly, caused an increase in intramammary and arterial blood pressure consistent with the release of both oxytocin and vasopressin; the release occurred irrespective of the state of the cortical EEG, and was blocked in every case by systemically applied atropine (0.1–1 mg/kg^{-1} i.v.). In contrast, the intravenous injection of mecamylamine (2 mg/kg^{-1}) had no effect upon the release of either hormone, though the responses evoked by carbachol were abolished in half of the animals when mecamylamine was injected intraventricularly (50 μg). When we caused a lesion to the SFO we also abolished the responses to carbachol.

Over 80% of the nonphasic and over 90% of the phasic neurons recorded in the PVN and SON were excited when 0.2 μg carbachol was injected intraventricularly. The latency to the onset of response was less than 30 s; the peak rate (on average a tenfold increase in firing rate) was reached within 2 min and lasted up to 30 min. In general, the phasic neurons changed first to a continuous period of discharge, which then broke up into a phasic pattern as the response declined (Fig. 9). The oxytocin neurons showed a rapid increase in the frequency of firing following the injection, then slowly and smoothly declined to the previous level of activity. Nonmagnocellular neurons in the vicinity of the SON were totally unaffected by the injection of carbachol.

The response of both phasic and nonphasic neurons to carbachol was totally abolished by systemic atropine (1 mg/kg^{-1}), though this antagonist had no effect on the spontaneous background activity. Similarly, atropine did not affect the neurosecretory bursts of oxytocin neurons during suckling, or the ability of the magnocellular neurons to respond to other excitants such as noradrenaline. A dose of the nicotinic antagonist mecamylamine (2 mg/kg^{-1} i.v.) which will block milk ejection during suckling (Clarke et al. 1978) only slightly reduced the response of the neurons to intraventricular carbachol, yet greatly reduced or totally suppressed the neurosecretory bursts of all the oxytocin neurons studied. The effects of injecting nicotine (20 μg) into the ventricles was also examined; within 30 s it produced an excitation which lasted 1–12 min, but in only half of the neurons studied. This excitation could be blocked by mecamylamine (2 mg/kg^{-1}) but not by atropine (1 mg/kg^{-1}).

The cholinoceptor antagonists atropine and mecamylamine were also studied on osmotic stimuli known to excite the magnocellular neurons. The excitation of both vasopressin and oxytocin neurons which followed the intraventricular injection of hypertonic sodium chloride (1 μl, 3*M*) was unaffected by intravenous mec-

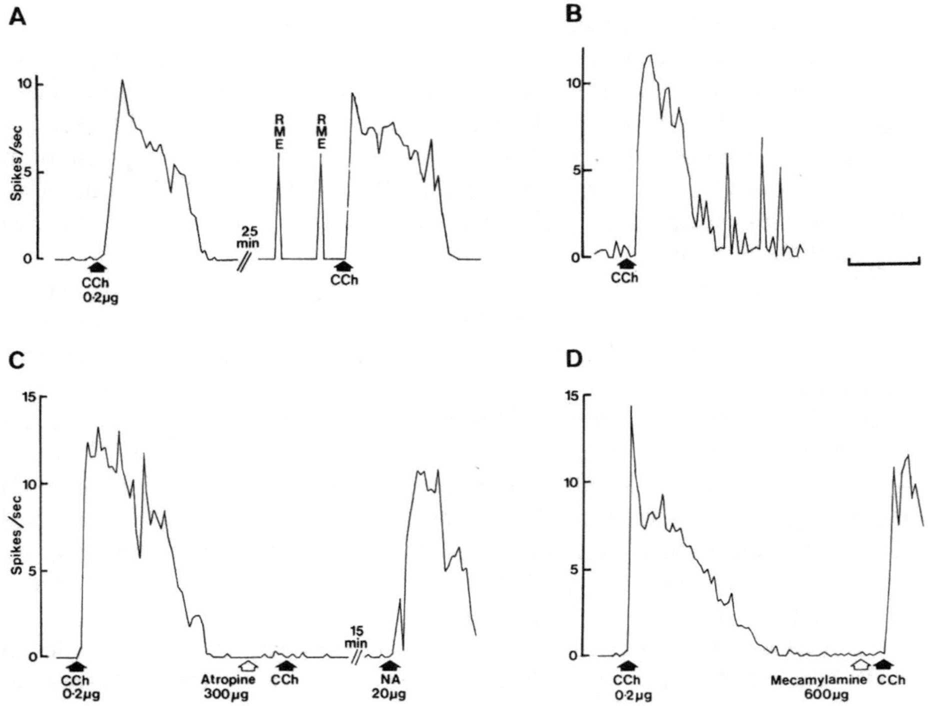

Fig. 9 a–d. The excitatory action of carbachol (*CCh*) on four supraoptic neurons. In each case the firing rate has been plotted against elapsed time. **a** The effect of 0.2 μg intraventricular carbachol (*black arrow*) on a quiescent oxytocin neuron; note the two neurosecretory bursts at reflex milk ejection (*RME*) prior to the second injection of carbachol. **b** Intraventricular carbachol exciting a phasic neuron. Note the phasic bursts towards the end of the response. **c** Following the initial response to carbachol, atropine 1 mg/kg^{-1} injected intravenously (*open arrow*) blocked a second response, though intraventricular noradrenaline (20 μg) still excited the neuron. **d** Mecamylamine (2 mg/kg^{-1}) failed to prevent an excitatory response to carbachol

amylamine or atropine; the excitatory responses were prevented in 84% of the neurons tested, however, by the intraventricular injection of mecamylamine (50 μg). The excitations evoked by haemorrhage were unaltered following atropine but were prevented by the systemic injection of mecamylamine (4 mg/kg^{-1}).

In summary, it would appear that vasopressin neurons have nicotinic cholinoceptors which are excitatory, whereas oxytocin neurons do not and may instead possess inhibitory muscarinic cholinoceptors. The excitation of both oxytocin and vasopressin neurons resulting from haemorrhage and of oxytocin neurons during suckling can be readily blocked by the nicotinic antagonist mecamylamine. Osmotic excitation is more resistant to blocking by cholinoceptor antagonists. Because of its muscarinic character the potent excitatory effect of carbachol injected into the ventricles cannot be due to a direct effect on the magnocellular neurons themselves. It would seem to be due to an action primarily on the SFO, a structure known to possess excitatory muscarinic receptors (Buranarugsa and Hubbard 1979) and to have connexions with the PVN and SON (Miselis 1981). Since under some circumstances both the osmotic- and the carbachol-evoked excitations could be blocked by a nicotinic antagonist (at a high dose) this projection may be cholinergic, as has been suggested previously (Bridges and Thorn 1970; Sladek and Joynt 1979).

4.5.2 Endogenous Opioid Peptides

Endogenous opioid peptides (for definition see Cox 1982) appear to be involved in the regulation of many endocrine systems, particularly at the level of the hypothalamus, and the hypothalamo-neurohypophyseal system seems to be no exception. Abundant immunohistological evidence indicates a close relationship between the magnocellular neurons and those showing strong immunoreactivity indicative of the presence of endogenous opioid peptides. Enkephalin- and dynorphin-like materials have been observed in perikarya within the SON and PVN; fibres are believed to project to the neurointermediate lobe, and in some cases the opioid peptides may even coexist with vasopressin (Rossier et al. 1979; Martin and Voigt 1981; Watson et al. 1982). It has been claimed that some neurons of the PVN may also have corticotropin-releasing factors coexisting with dynorphin and vasopressin (Roth et al. 1982), though these are probably neurons projecting to the median eminence. Adrenocorticotropin (ACTH) immunoreactive fibres of arcuate origin have also been located in the vicinity of the magnocellular neurons of the PVN and SON, particularly in association with the oxytocin neurons (Sawchenko et al. 1982). Therefore, β-endorphin may also be present, for β-endorphin and adrenocorticotropin are derived from the same precursor molecule, preproopiomelanocortin (O'Donohue and Dorsa 1982).

What may be the role, if any, of these opioid peptides in the regulation of vasopressin and oxytocin secretion? It has been known for many years, from clinical observations, that morphine and other opiates cause antidiuresis, and early experimental evidence suggested that this was due to a direct stimulatory action on vasopressin neurons (Duke et al. 1951). Ironically, it now seems that the more fundamental action of opiates and opioid peptides on magnocellular neurons is to produce a reduction in electrical activity and neurosecretion.

Application of enkephalin analogues at low doses to paraventricular neurons (a very heterogeneous population) in hypothalamic slices resulted in inhibition of activity of most neurons, including a few that showed phasic activity (Muehlethaler et al. 1980; Pittman et al. 1980). In our own experiments (Wakerley et al. 1983 a) we made recordings in vitro from neurons in the SON (an area where one can assume with rather more certainty that phasic neurons are hypothalamo-neurohypophyseal vasopressin neurons, and found that morphine and enkephalin analogues inhibited the activity of the continuously firing neurons, but that the activity of the phasic neurons was unaffected (Fig. 10). Similarly, when Arnauld et al. (1983) applied Leu-enkephalin by iontophoresis directly onto phasically firing supraoptic neurons in vivo they found it had only a weak inhibitory effect. One is tempted, therefore, to make the assumption that oxytocin neurons can be inhibited directly by opioid peptides but that vasopressin neurons cannot. Alternatively, it may be that the mechanism responsible for phasic discharge is

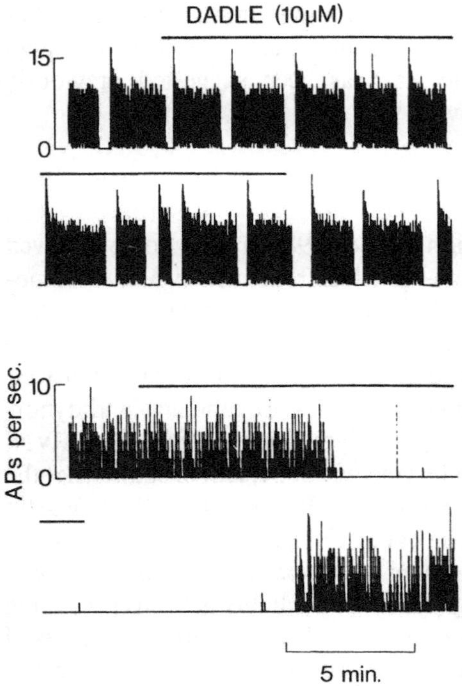

Fig. 10. Integrator records showing the effect of an opioid peptide on the activity of two neurons recorded in the supraoptic nucleus in an isolated rat hypothalamic slice. The upper two rows are a continuous record of a neuron discharging phasically; the lower two rows are of a neuron showing a continuous irregular discharge. For the period indicated by the *solid line* above the records, D-ALA-D-LEU-enkephalin (*DADLE*, a stable synthetic opioid) was added to the perfusing medium. Addition of enkephalin inhibited the continuous cell reversibly but had no effect upon the phasically firing neuron. Morphine (10 μ*M*) had similar actions. Over 60% of the nonphasic supraoptic neurons were inhibited, whereas less than 10% of the phasic neurons were affected. *AP*, action potentials (Figure supplied by Dr. J. B. Wakerley)

insensitive to opioids, while that for continuous activity can be modulated. Whatever the reason, it is of interest that the difference in sensitivity of the two types of neurons reflects to a certain extent the situation in the neural lobe (see Sect. 3.3.3).

A different profile of action is observed when one considers the effect of morphine and opioid peptides injected intraventricularly. In this type of experiment we found that all the phasically firing neurons were profoundly inhibited by morphine (2 μg i.c.v.) or β-endorphin (200 ng i.c.v.). The inhibition occurred within 10 s, lasted for over an hour, and could be reversed by the opiate antagonist naloxone (see Fig. 11). In contrast, the activity of the oxytocin neurons was little affected; some showed a small reduction in background activity and others were slightly excited. The neurosecretory bursts during suckling which precede milk ejection were entirely unaltered (Clarke et al. 1979 b). Our interpretation is that inhibition of phasic neurons by intraventricular injection of opioids is indirect, through reduction of an active afferent input such as the osmotic drive. As already discussed, material injected intraventricularly would not gain rapid access to the PVN and SON; again, the likely target would be periventricular structures which would be exposed to the injected material. It has previously been suggested that opioid peptides act on periventricular sodium-sensitive sites to reduce the drinking response and vasopressin release evoked by hypertonic sodium chloride and angiotensin (Summy-Long et al. 1981 a, b). We have observed that the administration of morphine or one of the opioid peptides prevents the excitation of the

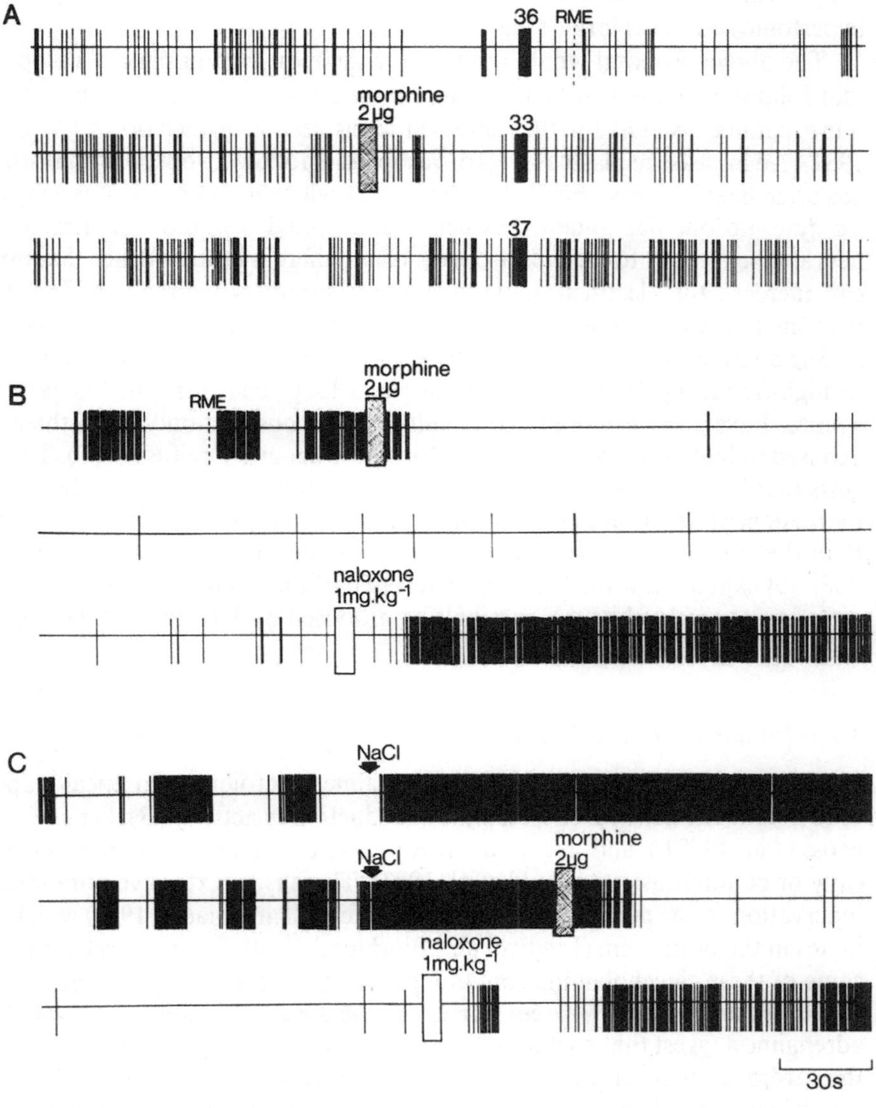

Fig. 11 a–c. The effect of injecting morphine (2 µg) intraventricularly on three identified magnocellular neurons. In each case the three rows are from a continuous record, but with approximately 5 min of the trace removed between each row. **a** An oxytocin neuron displayed neurosecretory bursts during suckling. Morphine had very little effect upon the ongoing activity and the magnitude of the burst (the *digits* give the number of potentials in the burst). Reflex milk ejection (*RME*) occurred prior to morphine administration, but ceased following the injection despite the occurrence of the neurosecretory burst. **b** A phasic neuron which did not respond prior to milk ejection (*RME*) was inhibited by intraventricular morphine; the activity returned following intravenous injection of naloxone (1 mg/kg^{-1}). **c** A phasic neuron was excited by intraventricular sodium chloride (1 µl, 3*M*); a second response to sodium chloride was terminated by the addition of morphine intraventricularly. Again, the opiate inhibition was reversed by 1 mg/kg^{-1} naloxone injected intravenously

magnocellular neurons which should result from the intraventricular injection of hypertonic sodium chloride (Fig. 11).

The above proposal would account for the reduction in plasma vasopressin that follows low doses of opioids, since most animals have some osmotic drive, but it cannot account for the increased levels seen at higher doses (Aziz et al. 1981). An increase in plasma levels of vasopressin following opiate administration has often been reported, but the mechanism by which it occurs remains a mystery. To date, no one has obtained evidence that opioid peptides can stimulate the neural lobe directly to secrete vasopressin, and there is little evidence that opiates can increase the electrical activity of the vasopressin neurons. A few of the neurons that we recorded (fewer than 10%) showed either an initial excitation, lasting a few seconds and followed by a long-lasting inhibition, or a short period of high-frequency discharge in the middle of long silent periods. Excitatory responses have also been observed in cultured neurons, but only when these were exposed to high doses of opioid peptides (Gahwiler and Dreifuss 1980). This suggests that it may be possible for vasopressin neurons to be excited either directly by these peptides or indirectly through an afferent input from brain areas other than the circumventricular organs. Even assuming that the neurons are excited does not explain how the hormone is released, if the opiates are also suppressing secretion at the level of the terminals as proposed by Iversen et al. (1980). This enigma has yet to be resolved.

4.5.3 Noradrenaline and Dopamine

There is general agreement that noradrenaline, microiontophoretically applied onto magnocellular neurons, results in a reduction in activity (Barker et al. 1971; Moss et al. 1972b), and this occurs irrespective of whether they are firing phasically or continuously (Arnauld et al. 1983). There is an extensive noradrenergic innervation to the magnocellular neurons (McNeill and Sladek 1980) which originates in the brain stem (Lindvall and Bjorklund 1974). The close relationship of some of these catecholamine-containing neurons to the solitary nucleus, the site of termination of vagal afferent fibres, and the observed inhibitory action of noradrenaline suggest that part of this ascending noradrenergic input might mediate the cardiovascular reflexes which inhibit vasopressin release. The inhibitory action of noradrenaline on the magnocellular neurons operates through β-adrenoceptors (Barker et al. 1971). A β inhibitory system also influences oxytocin release, though whether this operates at the level of the magnocellular neurons is debatable (Tribollet et al. 1978).

Another action of noradrenaline has also been reported, namely the stimulation of vasopressin release, or of physiological reflexes which cause it. Injection of noradrenaline into either the SON (Milton and Patterson 1974) or the ventricular system (Kuhn 1974) stimulated release of vasopressin which could be blocked by α-adrenoceptor antagonists. These same antagonists have been reported to prevent the release of vasopressin by osmotic stimulation (Bridges and Thorn 1970) and vagal stimulation (Mills and Wang 1964). Conversely, an α-mediated inhibitory action on vasopressin release has been proposed (Kimura et al. 1981; Armstrong et al. 1982). In our own experience of recording from magnocellular neurons, they have been excited by the α-adrenoceptor agonist phenylephrine and by noradrenaline injected intraventricularly in vivo (see Fig. 9), and when added to the perfusion medium bathing the hypothalamic slice in vitro (Wakerley et al. 1983 b).

Experiments involving dopamine, or agonists and antagonists which act at dopamine receptors, have been equally confusing. Using an isolated hypothalamo-neurohypophyseal system, Bridges et al. (1976) proposed that dopamine and apomorphine stimulated vasopressin and oxytocin release. In contrast, Seybold and colleagues (1978) – using a similar preparation – claimed that dopamine and apomorphine inhibited the release of vasopressin. Both groups suggested that the action was at the level of the hypothalamus (see also Sect. 3.3.2). Injection of dopamine or dopamine mimetics into the ventricles results in the release of both vasopressin and oxytocin (Bridges et al. 1976; Clarke et al. 1979 a; Moos and Richard 1979; Kimura et al. 1981). Moos and Richard (1982) examined the activity of magnocellular neurons in the PVN and concluded that dopamine has a facilitative role in exciting afferent pathways to both oxytocin and vasopressin neurons. We have also observed excitation of the magnocellular neurons in the SON following intraventricular administration of dopamine and apomorphine. The only microiontophoretic study in which dopamine has been applied to magnocellular neurons gave equivocal results: half of the neurons were excited and half inhibited (Moss et al. 1972 a). It is likely that the varied conclusions drawn from the evidence regarding catecholamine involvement in neurohypophyseal hormone release merely reflect the complexity of central noradrenergic and dopaminergic systems. In order to resolve these problems it will be necessary to take into account the multiplicity of receptors and the several potential mechanisms of action displayed by these catecholamines.

4.5.4 Vasopressin and Oxytocin

In 1972, Dreifuss and Kelly (1972 b) suggested that the suppression of activity which followed antidromic stimulation of magnocellular neurons was due to recurrent collaterals. This implies that the neurohypophyseal hormones vasopressin and oxytocin may act as neurotransmitters. This view has gained further support from the histological evidence that nerve fibres containing vasopressin- and oxytocin-like immunoreactivity project to many parts of the brain other than the posterior pituitary. Although there is increasing evidence that these neurons may also contain other neuroactive peptides, as discussed in Sect. 3.2, the possibility that the neurohypophyseal peptides act as neurotransmitters or modulators cannot be excluded. Certainly, both vasopressin and oxytocin injected into the cerebral ventricles or specific brain areas have profound effects upon a variety of brain processes (van Ree et al. 1978; Kasting et al. 1979; Urban 1981; Freund-Mercier and Richard 1981). The neurohypophyseal hormones are released in several brain regions in a way analogous to that of other neurotransmitters (Buijs and van Heerikhuize 1982), and several physiological stimuli release the peptides into the CSF (Barnard and Morris 1982; Robinson and Jones 1982; Wang et al. 1982). In the forebrain iontophoretically applied vasopressin and oxytocin excited septal and hippocampal neurons (Joels and Urban 1982), as did vasopressin when bath-applied to the hippocampal slice (Muehlethaler and Dreifuss 1982). Neurons in the brain stem and spinal cord, however, tend to be inhibited by the neurohypophyseal peptides applied microiontophoretically (Morris et al. 1980; Gilbey et al. 1982; Harris et al. 1982). As for the magnocellular neurons, vasopressin has been reported to have a weak inhibitory action on them when applied iontophoretically (Nicoll and Barker 1971), whereas oxytocin excited them (Moss et al. 1972a). Thus, the basis for recurrent inhibition through the action of the neurohypophyseal peptides exists, by vasopressin acting directly on the magnocellular neurons or by either vasopressin or oxytocin acting indirectly. The inhibitory response following antidromic stimulation is certainly impaired in Brattleboro rats deficient in vasopressin (Leng and Wiersma 1981); the exact involvement of the neurohypophyseal peptides, however, is still subject to debate (Leng 1981).

5 Summary

The magnocellular neurons of the hypothalamus, though somewhat unusual in that they synthesize peptides which they release into the circulation, appear to possess many of the electrophysiological characteristics of neurons in other areas of the brain. They receive synaptic input which causes excitatory and inhibitory post-synaptic potentials, leading to the generation of typical action potentials. These action potentials are propagated along unmyelinated fibres to the terminals to cause the release of the neurosecretory product by calcium-mediated exocytosis. The overall rate of discharge of the neurons is low, except when they are being driven by stimuli which are function specific. The level of electrical activity within the neurons relates directly to the quantity of hormone secreted, and when driven strongly the neurons show characteristic patterns of activity; the vasopressin neurons show phasic activity and the oxytocin neurons, during parturition and suckling, give high-frequency bursts of activity related to hormone release. Use of experimental stimuli which mimic these characteristic patterns of discharge has revealed that they facilitate the release process at the level of the nerve terminals. In general, the ongoing activity of the magnocellular neurons appears to be independent of the level of arousal of the animal, though there may be subtle changes of the temporal spacing of the action potentials coincident with changes in the cortical EEG.

Several powerful and reasonably specific stimuli influence the activity of the magnocellular neurons and have been studied in detail. The most specific stimulus known to date is that of the young sucking on the nipples during nursing; this results in characteristic activation of the oxytocin neurons exclusively, and has been used extensively as a means of distinguishing between oxytocin- and vasopressin-containing neurons. The oxytocin neurons display a similar type of activity during parturition, though many vasopressin neurons are also excited during labour. Bilateral carotid occlusion has been claimed to be a specific stimulus for activating vasopressin neurons; however, most hypovolaemic stimuli result in the excitation of both vasopressin and oxytocin neurons. Conversely, stimuli reflecting an elevated blood volume (hypervolaemia) cause a reduction in the activity of the magnocellular neurons. An afferent pathway has been identified – projecting from the medulla to the magnocellular nuclei – which may be monosynaptic; reciprocal vasopressin- and oxytocin-containing fibres are present.

One of the most effective and functionally active stimuli for the release of vasopressin is that of osmotic stimulation resulting from either dehydration or osmotic loading. Curiously, whatever form the osmotic stimulus takes, both the oxytocin neurons and the vasopressin neurons are excited. The magnocellular neurons themselves are probably osmosensitive, yet synaptic input from remote osmoreceptors appears to be important for the full expression of an osmotic response. Although there are probably several sites of osmosensitivity, the circumventricular organs are undoubtedly involved; they respond to elevated osmolality of blood and CSF, but the sodium ion concentration seems to be the most important factor.

Many neurotransmitters have been implicated in the regulation of the magnocellular neurons but acetylcholine, noradrenaline and the opioid peptides have been studied most. A readily antagonized cholinergic (nicotinic) component forms part of the neural pathway involved in the excitatory hypovolaemic stimulus and in the afferent limb of the milk-ejection reflex. These nicotinic excitatory receptors may even be located on the magnocells in the case of the vasopressin neurons. A cholinergic (muscarinic) component is also responsible for excitation of the magnocellular neurons, though this appears to be located in the circumventricular organs. Noradrenaline also excites magnocellular neurons and releases vasopressin and oxytocin, though the action is probably indirect, through α-adrenoceptors. Noradrenaline acting directly on the magnocellular neurons operates through β-adrenoceptors to inhibit cell activity. Aminergic mechanisms are implicated in the haemodynamic, osmotic and suckling-evoked release of neurohypophyseal hormones. Opioid peptides inhibit the activity of the magnocellular neurons indirectly by acting on the circumventricular organs to suppress the osmotic drive; they may also directly inhibit the oxytocin neurons, but a physiologically related action at this site has yet to be established.

One of the most important and fundamental observations that has evolved from electrophysiological studies of the magnocellular neurons is the correlation of their electrical activity with the release of the neurohypophyseal peptides. Even this relationship may be subject to modification, however, for it has been shown that other neural mechanisms may operate in the neurointermediate lobe to influence the terminals of the magnocellular neurons and modulate the release of vasopressin and oxytocin. Noradrenaline, dopamine and the opioid peptides have all been shown to act on the neural lobe, in some cases facilitating and in others inhibiting the release of the neurohypophyseal hormones. At present, however, the only commanding evidence for a physiological and functional role is that for opioid peptides inhibiting the release of oxytocin.

Acknowledgement. We wish to thank our colleagues for their help and guidance during the preparation of this chapter, in particular Karen Jukes for her limitless patience and expert technical assistance.

We are grateful to the Medical Research Council of Great Britain for its financial support.

References

Ahn MS, Feldman SC, Makman MH (1979) Posterior pituitary adenylate cyclase: stimulation by dopamine and other agents. Brain Res 166:422–425

Akaishi T, Negoro H, Kobayasi S (1980) Responses of paraventricular and supraoptic units to angiotensin II, sar^1-ile^3-angiotensin II, and hypertonic NaCl administered into the cerebral ventricle. Brain Res 188:499–511

Alper RH, Moore KE (1982) Injection of hypertonic saline or mannitol accelerates the dehydration-induced activation of dopamine synthesis in the neurointermediate lobe of the rat hypophysis. Neuroendocrinology 35:469–474

Ambach G, Palkovits M (1979) The blood supply of the hypothalamus of the rat. Morgane PJ, Panksepp J (eds) In: Anatomy of the hypothalamus. Dekker, New York

Andersson B (1978) Regulation of water intake. Physiol Rev 58:582–603

Armstrong WE, Sladek CD, Sladek JR (1982) Characterization of noradrenergic control of vasopressin release by the organ-cultured rat hypothalamo-neurohypophyseal system. Endocrinology 111:273–9

Arnauld E, Cirino M, Layton BS, Renaud LP (1983) Contrasting actions of amino acids, acetylcholine, noradrenaline and leucine enkephalin on the excitability of supraoptic vasopressin-secreting neurones. Neuroendocrinology 36:187–196

Aziz LA, Forsling ML, Woolf CJ (1981) The effect of intracerebroventricular injections of morphine on vasopressin release in the rat. J Physiol (Lond) 311:401–409

Balment RJ, Brimble MJ, Forsling ML (1980) Release of oxytocin induced by salt loading and its influence on renal excretion in the male rat. J Physiol (Lond) 308:439-449

Barden N, Chevillard C, Saavedra JM (1982) Diurnal variations in rat posterior pituitary catecholamine levels. Neuroendocrinology 34:148–150

Barker JL, Crayton JW, Nicoll RA (1971) Noradrenaline and acetylcholine responses of supraoptic neurosecretory cells. J Physiol (Lond) 218:19–32

Barnard RR, Morris M (1982) Cerebrospinal fluid vasopressin and oxytocin: evidence for an osmotic response. Neurosci Lett 29:275–279

Bennett CT, Pert A (1974) Antidiuresis produced by injections of histamine into the cat supraoptic nucleus. Brain Res 78:151–156

Bicknell RJ, Leng G (1982) Endogenous opiates regulate oxytocin but not vasopressin secretion from the neurohypophysis. Nature 298:161–162

Bie P (1980) Osmoreceptors, vasopressin and control of renal water excretion. Physiol Rev 60:961–1048

Bioulac B, Gaffori O, Harris M, Vincent J-D (1978) Effects of acetylcholine, sodium glutamate and GABA on the discharge of supraoptic neurons in the rat. Brain Res 154:159–162

Bisset GW, Chowdrey HS (1981) A central cholinergic link in the neural control of the release of vasopressin. Br J Pharmacol 74:239

Bisset GW, Feldberg W, Guertzenstein PG, Rocha E, Silva M Jr (1975) Vasopressin release by nicotine: the site of action. Br J Pharmacol 54:463–474

Blume HW, Pittman QJ, Renaud LP (1978) Electrophysiological indications of a "vasopressinergic" innervation of the median eminence. Brain Res 155:153–159

Bridges TE, Hillhouse EW, Jones MT (1976) The effect of dopamine on neurohypophysial hormone release in vivo and from the rat neural lobe and hypothalamus in vitro. J Physiol (Lond) 260:647–666

Bridges TE, Thorn NA (1970) The effect of autonomic blocking agents on vasopressin release in vivo induced by osmoreceptor stimulation. J Endocrinol 48:265–276

Brimble MJ, Dyball REJ (1977) Characterization of the responses of oxytocin- and vasopressin-secreting neurones in the supraoptic nucleus to osmotic stimulation. J Physiol 271:253–271

Brimble MJ, Dyball REJ, Forsling ML (1978) Oxytocin release following osmotic activation of oxytocin neurones in the paraventricular and supraoptic nuclei. J Physiol 278:69–78

Buggy J, Hoffman WE, Phillips MI, Fisher AE, Johnson AK (1979) Osmosensitivity of rat third ventricle and interactions with angiotensin. Am Physiol Soc 236:R75–82

Buijs RM, Swaab DF, Dogterom J, van Leeuwen FW (1978) Intra- and extrahypothalamic vasopressin and oxytocin pathways in the rat. Cell Tissue Res 186:423–433

Buijs RM, van Heerikhuize JJ (1982) Vasopressin and oxytocin release in the brain – a synaptic event. Brain Res 252:71–76

Buranarugsa P, Hubbard JI (1979) The neuronal organization of the rat subfornical organ in vitro and a test of the osmo- and morphine-receptor hypotheses. J Physiol 291:101–116

Chapman DB, Way EL (1980) Metal ion interactions with opiates. Annu Rev Pharmacol Toxicol 20:553–579

Christensen JD, Fjalland B (1982) Lack of effect of opiates on release of vasopressin from isolated rat neurohypophysis. Acta Pharmacol Toxicol (Copenh) 50:113–116

Ciriello J, Calaresu FR (1980) Monosynaptic pathway from cardiovascular neurons in the nucleus tractus solitarii to the paraventricular nucleus in the cat. Brain Res 193:529–533

Clarke G, Merrick LP (1982) Stimulation of vasopressin and oxytocin neurones by hypertonic sodium chloride injected into the cerebral ventricles. J Physiol (Lond) 327:24–25

Clarke G, Patrick G (1983) Differential inhibitory action by morphine on the release of oxytocin and vasopressin from isolated neural lobe. Neurosci Lett 39:175–180

Clarke G, Fall CHD, Lincoln DW, Merrick LP (1978) Effects of cholinoceptor antagonists on the suckling-induced and experimentally evoked release of oxytocin. Br J Pharmacol 63:519–527

Clarke G, Lincoln DW, Merrick LP (1979 a) Dopaminergic control of oxytocin release in lactating rats. J Endocrinol 83:409–420

Clarke G, Wood P, Merrick L, Lincoln DW (1979 b) Opiate inhibition of peptide release from the neurohumoral terminals of hypothalamic neurones. Nature 282:746–748

Clarke G, Kirby PJC, Thomson AM (1980) Effects on vasopressinergic and oxytocinergic neurones of intraventricular substance P. J Physiol (Lond) 307:59

Cowley AW (1982) Vasopressin and cardiovascular regulation. Int Rev Physiol 26:189–242

Cox BM (1982) Mini review: endogenous opioid peptides: a guide to structures and terminology. Life Sci 31:1645–1658

Dellman HD, Simpson JB (1979) The subfornical organ. Int Rev Cytol 58:333–421

Douglas WW (1974) Mechanism of release of neurohypophysial hormones: stimulus-secretion coupling. In: Handbook of physiology-endocrinology. American Physiological Society, Washington D. C. Vol IV, Part 1, pp 191–224

Douglas WW, Poisner AM (1964) Stimulus-secretion coupling in a neurosecretory organ and the role of calcium in the release of vasopressin from the neurohypophysis. J Physiol (Lond) 172:1–18

Dreifuss JJ, Kelly JS (1972 a) The activity of identified supraoptic neurones and their response to acetylcholine applied by iontophoresis. J Physiol (Lond) 220:105–118

Dreifuss JJ, Kelly JS (1972 b) Recurrent inhibition of antidromically identified rat supraoptic neurones. J Physiol (Lond) 220:87–103

Dreifuss JJ, Kalnins I, Kelly JS, Ruf KB (1971) Action potentials and release of neurohypophysial hormones in vitro. J Physiol 215:805–817

Dreifuss JJ, Harris MC, Tribollet E (1976) Excitation of phasically firing hypothalamic supraoptic neurones by carotid occlusion in rats. J Physiol (Lond) 257:337–354

Dreifuss JJ, Tribollet E, Baertschi AJ (1976) Excitation of supraoptic neurones by vaginal distention in lactating rats; correlation with neurohypophysial hormone release. Brain Res 113:600–605

Dreifuss JJ, Tribollet E, Muhlethaler (1981) Temporal patterns of neural activity and their relation to the secretion of posterior pituitary hormones. Biol Reprod 24:51–72

Duke HN, Pickford M, Watt JA (1951) The antidiuretic action of morphine, its site and mode of action in the hypothalamus of the dog. Q J Exp Physiol 36:149–158

Dutton A, Dyball REJ (1979) Phasic firing enhances vasopressin release from the rat neurohypophysis. J Physiol 290:433–440

Dyball REJ (1971) Oxytocin and ADH secretion in relation to eletrical activity in antidromically identified supraoptic and paraventricular units. J Physiol 214:245–256

Dyball REJ, McPhaill CI (1974) Unit activity in the supraoptic and paraventricular nuclei – the effects of anaesthetics. Brain Res 67:43–50
Dyball REJ, Prilusky J (1981) Responses of supraoptic neurones in the intact and deafferented rat hypothalamus to injections of hypertonic sodium chloride. J Physiol (Lond) 311:443–452
Fisher AWF, Price PG, Burford GD, Lederis K (1979) A 3-dimensional reconstruction of the hypothalamo-neurohypophysial system of the rat. Cell Tissue Res 204:343–354
Freund-Mercier MJ, Richard Ph (1981) Excitatory effects of intraventricular injections of oxytocin on the milk-ejection reflex in the rat. Neurosci Lett 23:193–198
Gahwiler BH, Dreifuss JJ (1979) Phasically firing neurons in long-term cultures of the rat hypothalamic supraoptic area: pacemaker and follower cells. Brain Res 177:95–103
Gahwiler BH, Dreifuss JJ (1980) Transition from random to phasic firing induced in neurons cultured from the hypothalamic supraoptic area. Brain Res 193:415–425
Ganten D, Lang RE, Lehmann E, Unger Th (1984) Brain angiotensin: On the way to becoming a well-studied neuropeptide system. Biochem Pharmacol 33:3523–3528
Gauer OH, Henry JP, Behn C (1970) The regulation of extracellular fluid volume. Annu Rev Physiol 32:547–595
Gilbey MP, Coote JH, Fleetwood-Walker S, Peterson DF (1982) The influence of the paraventriculo-spinal pathway and oxytocin and vasopressin on sympathetic preganglionic neurons. Brain Res 251:283–290
Haas HL, Wolf P, Nussbaumer J-C (1975) Histamine action on supraoptic and other hypothalamic neurones of the cat. Brain Res 88:166–170
Harris GW, Manabe Y, Ruf KB (1969) A study of the parameters of electrical stimulation of unmyelinated fibres in the pituitary stalk. J Physiol (Lond) 203:67–81
Harris MC (1979) Effects of chemoreceptor and baroreceptor stimulation on the discharge of hypothalamic supraoptic neurones in rats. J Endocrinol 82:115–125
Harris MC, Banks D, Zerihun L (1982) Inputs from hypothalamic paraventricular nucleus to dorsal medullary nuclei in the rat. Baertschi AJ, Dreifuss JJ (eds) In: Neuroendocrinology of vasopressin corticoliberin and opiomelanocortin. Academic, London, pp 153–165
Harris MC, Dreifuss JJ, Legross JJ (1975) Excitation of phasically firing supraoptic neurones during vasopressin release. Nature 258:80–82
Hatton GI (1982) Phasic bursting activity of rat paraventricular neurones in the absence of synaptic transmission. J Physiol (Lond) 327:273–284
Hatton GI, Armstrong WE, Gregory WA (1978) Spontaneous and osmotically stimulated activity in slices of rat hypothalamus. Brain Res Bull 3:497–508
Hayward JN, Jennings DP (1973) Activity of magnocellular neuroendocrine cells in the hypothalamus of unanaesthetised monkeys. 1. Functional cell types and their anatomical distribution in the supraoptiv nucleus and internuclear zone. J Physiol 232:515–543
Hoffman PK, Share L, Crofton JT, Shade RE (1982) The effect of intracerebroventricular indomethacin on osmotically stimulated vasopressin release. Neuroendocrinology 34:132–139
Hoffman WE, Schmid PG (1979) Cardiovascular and antidiuretic effects of central prostaglandin E_2. J Physiol (Lond) 288:159–169
Hoffman WE, Phillips MI, Schmid P (1977) The role of catecholamines in central antidiuretic and pressor mechanisms. Neuropharmacology 16:563–569
Holzbauer M, Sharman DF, Godden U (1978) Observations on the function of the dopaminergic nerves innervating the pituitary gland. Neuroscience 3:1251–1262
Ingram CD, Bicknell RJ, Brown D, Leng G (1982) Rapid fatigue of neuropeptide secretion during continual electrical stimulation. Neuroendocrinology 35:424–428
Iversen LL, Iversen SD, Bloom FE (1980) Opiate receptors control vasopressin release from nerve terminals in rat neurohypophysis. Nature 284:350–351
Jennings DP, Haskins JT, Rogers JM (1978) Comparison of firing patterns and sensory responsiveness between supraoptic and other hypothalamic neurons in the unanesthetized sheep. Brain Res 149:347–364
Joels M, Urban IJA (1982) The effect of microiontophoretically applied vasopressin and oxytocin on single neurones in the septum and dorsal hippocampus of the rat. Neurosci Lett 33:79–84

Karasek M, Traczyk WZ, Orlowska-Majdak M, Rubacha G (1980) Quantitative changes of the neurosecretory granules in the rat neurohypophysis after preganglionic stimulation of the superior cervical ganglion. Acta Med Pol 21:351–352
Kasting NW, Cooper KE, Veale WL (1979) Antipyresis following perfusion of brain site with vasopressin. Experientia 35:208–209
Keil LC, Summy-Long J, Severs WB (1975) Release of vasopressin by angiotensin II. Endocrinology 96:1063–1065
Kimura T, Share L, Wang BC, Crofton JT (1981) Central effects of dopamine and bromocriptine on vasopressin release and blood pressure. Neuroendocrinology 33:347–351
Koizumi K, Yamashita H (1978) Influence of atrial stretch receptors on hypothalamic neurosecretory neurones. J Physiol (Lond) 285:341–358
Kuhn ER (1974) Cholinergic and adrenergic release mechanism for vasopressin in the male rat: a study with injections of neurotransmitters and blocking agents into the third ventricle. Neuroendocrinology 16:255–264
Kuhn ER, McCann SM (1971) Release of oxytocin and vasopressin in lactating rats after injection of carbachol into the third ventricle. Neuroendocrinology 8:48–58
Leng G (1980) Rat supraoptic neurones: the effects of locally applied hypertonic saline. J Physiol (Lond) 304:405–414
Leng G (1981) The effects of neural-stalk stimulation upon firing patterns in rat supraoptic neurones. Exp Brain Res 41:135–145
Leng G, Mason WT, Dyer RG (1982) The supraoptic nucleus as an osmoreceptor. Neuroendocrinology 34:75–82
Leng G, Wiersma J (1981) Effects of neural-stalk stimulation on phasic discharge of supraoptic neurones in Brattleboro rats devoid of vasopressin. J Endocrinol 90:211–220
Lightman SL, Iversen LI, Forsling MI (1982) Dopamine and (D-Ala, D-Leu) enkephalin inhibit the electrically stimulated neurohypophyseal release of vasopressin in vitro: evidence for calcium-dependent opiate action. J Neurosci 2:78–81
Lincoln DW (1969) Correlation of unit activity in the hypothalamus with EEG patterns associated with the sleep cycle. Exp Neurol 24:1–18
Lincoln DW (1974) Dynamics of oxytocin secretion. In: Knowls F, Vollrath L (eds) Neurosecretion – the final neuroendocrine pathway. Springer, Berlin Heidelberg New York
Lincoln DW, Wakerley JB (1974) Electrophysiological evidence for the activation of supraoptic neurones during the release of oxytocin. J Physiol 242:533–554
Lincoln DW, Wakerley JB (1975) Factors governing the periodic activation of supraoptic and paraventricular neurosecretory cells during suckling in the rat. J Physiol 250:443–461
Lincoln DW, Hill A, Wakerley JB (1973) The milk-ejection reflex of the rat: an intermittent function not abolished by surgical levels of anaesthesia. J Endocrinol 57:459–476
Lincoln DW, Clarke G, Mason CA, Dreifuss JJ (1976) Physiological mechanisms determining the release of oxytocin in milk ejection and labour. In: Moses AH, Shave L (eds) neurohypophysis. Karger, Basel
Lincoln DW, Hentzen K, Hin T, Shoot P, Clarke G, Summerlee AJS (1980) Sleep: a prerequisite for reflex milk ejection in the rat. Exp Brain Res 38:151–162
Lind RW, Swanson LW, Ganten D (1985) Organization of angiotensin II immunoreactive cells and fibers in the rat central nervous system. Neuroendocrinology 40:2–24
Lindvall O, Bjorklund A (1974) The organization of the ascending catecholamine neuron systems in the rat brain. Acta Physiol Scand [Suppl] 314
Martin R, Voigt KH (1981) Enkephalins co-exist with oxytocin and vasopressin in nerve terminals of rat neurohypophysis. Nature 289:502–504
Mason WT (1980) Supraoptic neurones of rat hypothalamus are osmosensitive. Nature 287:154–157
McNeill TH, Sladek JR Jr (1980) Simultaneous monoamine histofluorescence and neuropeptide immunocytochemistry. II. Correlative distribution of catecholamine varicosities and magnocellular neurosecretory neurons in the rat supraoptic and paraventricular nuclei. J Comp Neurol 193:1023–1033
Menninger RP (1979) Effects of carotid occlusion and left atrial stretch on supraoptic neurosecretory cells. Am J Physiol 237:R63–67

Meyer DK, Phillips MI, Eiden L (1982) Studies on the presence of angiotensin II in rat brain. J Neurochem 38:816–820
Mills E, Wang SC (1964) Liberation of ADH: Pharmacologic blockade of ascending pathways. Am J Physiol 207:1405–1410
Milton AS, Paterson AT (1974) A microinjection study of the control of antidiuretic hormone release by the supraoptic nucleus of the hypothalamus in the cat. J Physiol (Lond) 241:607–628
Miselis RR (1981) The efferent projections of the subfornical organ of the rat: a circumventricular organ within a neural network subserving water balance. Brain Res 230:1–23
Moos F, Richard PH (1979) Effects of dopaminergic antagonist and agonist on oxytocin release induced by various stimuli. Neuroendocrinology 28:138–144
Moos F, Richard P (1982) Excitatory effect of dopamine on oxytocin and vasopressin reflex releases in the rat. Brain Res 241:249–260
Moos F, Richard P (1983) Serotonergic control of oxytocin release during suckling in the rat: opposite effects in conscious and anaesthetized rats. Neuroendocrinology 36:300–306
Morris JF, Nordmann JJ, Shaw FP (1981) Granules, microvesicles, and vacuoles – their roles in the functional compartments of the neural lobe. In: Farner DS, Lederis K (eds) Neurosecretion. Plenum, New York
Morris R, Salt TE, Sofroniew MV, Hill RG (1980) Actions of microiontophoretically applied oxytocin and immunohistochemical localisation of oxytocin vasopressin and neurophysin in the rat caudal medulla. Neurosci Lett 18:163–168
Moss RL, Dyball REJ, Cross BA (1972a) Excitation of antidromically identified neurosecretory cells of the paraventricular nucleus by oxytocin applied iontophoretically. Exp Neurol 34:95–102
Moss RL, Urban I, Cross BA (1972b) Microelectrophoresis of cholinergic and aminergic drugs on paraventricular neurons. Am J Physiol 223:310–318
Muehlethaler M, Dreifuss JJ (1982) A neuronal action of posterior pituitary peptides in the rat hippocampus. In: Neuroendocrinology of vasopressin, corticoliberin and opiomelanocortins. Academic, London
Muehlethaler M, Gaehwiler BH, Dreifuss JJ (1980) Enkephalin-induced inhibition of hypothalamic paraventricular neurones. Brain Res 197:264–268
Neve HA, Paisley AC, Summerlee AJ (1982) Arousal a pre-requisite for suckling in the conscious rabbit? Physiol Behav 28:213–217
Nicoll RA, Barker JL (1971) The pharmacology of recurrent inhibition in the supraoptic neurosecretory system. Brain Res 35:501–511
Nilaver G, Zimmerman EA, Wilkins J, Michaels J, Hoffman D, Silverman AJ (1980) Magnocellular hypothalamic projections to the lower brain stem and spinal cord of the rat. Neuroendocrinology 30:150–158
Nordmann JJ, Chevallier J (1980) The role of microvesicles in buffering [Ca^{2+}]: in the neurohypophysis. Nature 287:54–56
Nordmann JJ, Dreifuss JJ (1972) Hormone release evoked by electrical stimulation of rat neurohypophyses in the absence of action potentials. Brain Res 45:604–607
O'Donohue TL, Dorsa DM (1982) The opiomelanotropinergic neuronal and endocrine systems. Peptides 3:353–395
Oertel WH, Muenaini E, Tappaz ML, Weise VK, Dahl AL, Schmechel DE, Kopin IJ (1982) Central GABAergic innervation of neurointermediate pituitary lobe: biochemical and immunocytochemical study in the rat. Proc Natl Acad Sci USA 79:675–679
Pickford M (1939) The inhibitory effect of acetylcholine on water diuresis in the dog and its pituitary transmission. J Physiol 95:226–238
Pickford M (1947) The action of acetylcholine in the supraoptic nucleus of the chloralosed dog. J Physiol (London) 106:264–270
Pittman QJ, Hatton JD, Bloom FE (1980) Morphine and opioid peptides reduce paraventricular neuronal activity: studies on the rat hypothalamic slice preparation. Proc Natl Acad Sci USA 77:5527–5531
Pittman QJ, Veale WL, Lederis K (1982) Central neurohypophyseal peptide pathways: interactions with endocrine and other autonomic functions. Peptides 3:515–520
Poulain DA, Wakerley JB (1982) Electrophysiology of hypothalamic magnocellular neurones secreting oxytocin and vasopressin. Neuroscience 7:773–808

Poulain DA, Wakerley JB, Dyball REJ (1977) Electrophysiological differentiation of oxytocin- and vasopressin-secreting neurones. Proc R Soc Lond [Biol] 196:367–384

Poulain DA, Ellendorff F, Vincent JD (1980) Septal connections with identified oxytocin and vasopressin neurones in the supraoptic nucleus of the rat. An electrophysiological investigation. Neuroscience 5:379–387

Poulain DA, Rodriguez F, Ellendorff F (1981 a) Sleep is not a prerequisite for the milk-ejection reflex in the pig. Exp Brain Res 43:107–110

Poulain DA, Lebrun CJ, Vincent JD (1981 b) Electrophysiological evidence for connections between septal neurones and the supraoptic nucleus of the hypothalamus of the rat. Exp Brain Res 42:260–268

Racké K, Ritzel H, Trapp B, Muschell E (1982 a) Dopaminergic modulation evoked vasopressin release from the isolated neurohypophysis of the rat. Naunyn Schmiedebergs Arch Pharmacol 319:56–65

Racké K, Rothländer M, Muscholl E (1982 b) Isoprenaline and forskolin increase evoked vasopressin release from rat pituitary. Eur J Pharmacol 82:97–100

Reid IA (1979) The brain renin-angiotensin system: a critical analysis. Fed Proc 38:2255–2259

Reid IA, Brooks VL, Rudolph CD, Keil LC (1982) Analysis of the actions of angiotensin on the central nervous system of conscious dogs. Am J Physiol 243:R82–91

Rhodes CH, Morrell JI, Pfaff DW (1981 a) Immunohistochemical analysis of magnocellular elements in rat hypothalamus: distribution and numbers of cells containing neurophysin, oxytocin and vasopressin. J Comp Neurol 198:45–64

Rhodes CH, Morrell JI, Pfaff DW (1981 b) Distribution of estrogen-concentrating, neurophysin-containing magnocellular neurones in the rat hypothalamus as demonstrated by a technique combining steroid autoradiography and immunohistology in the same tissue. Neuroendocrinology 33:18–23

Robinson ICAF, Jones PM (1982) Oxytocin and neurophysin in plasma and CSF during suckling in the guinea-pig. Neuroendocrinology 34:59–63

Rossier J, Battenberg E, Pittman Q, Bayon A, Koda L, Miller R, Guillemin R, Bloom F (1979) Hypothalamic enkephalin neurones may regulate the neurohypophysis. Nature 277:653–655

Roth KA, Weber E, Barchas JD, Chang D, Chang J-W (1982) Immunoreactive dynorphin-(1-8) and corticotropin releasing factor in subpopulation of hypothalamic neurones. Science 219:189–191

Saavedra JM, Palkovitz M, Kizer JS, Brownstein M, Zivin JA (1975) Distribution of biogenic amines and related enzymes in the rat pituitary gland. J Neurochem 25:257–260

Sawchenko PE, Swanson LW, Joseph SA (1982) The distribution and cells of origin of ACTH (1-39)-stained varicosities in the paraventricular and supraoptic nuclei. Brain Res 232:365–374

Seybold VS, Miller JW, Lewis PR (1978) Investigation of a dopaminergic mechanism for regulating oxytocin release. J Pharmacol Exp Ther 207:605–610

Share L (1979) Interrelations between vasopressin and the renin-angiotensin system. Fed Proc 38:2267–2271

Share L, Levy MN (1966) Carotid sinus pulse pressure, a determinant of plasma and antidiuretic hormone concentration. Am J Physiol 211:721–724

Sharman DC, Holzer P, Holzbauer M (1982) In vitro release of endogenous catecholamines from the neural and intermediate lobe of the hypophysis. Neuroendocrinology 34:175–179

Shaw FD, Morris JF (1980) Calcium localization in the rat neurophypophysis. Nature 287:56–58

Silverman AJ, Hoffman DL, Zimmerman EA (1981) The descending afferent connections of the paraventricular nucleus of the hypothalamus (PVN). Brain Res Bull 6:47–61

Simantov R, Snyder SH (1977) Opiate receptor binding in the pituitary gland. Brain Res 124:178–184

Simonnet G, Bioulac B, Rodriguez F, Vincent JD 1980) Evidence of a direct action of angiotensin II on neurones in the septum and in the medial preoptic area. Pharmacol Biochem Behav 13:359–363

Simpson JB (1981) The circumventricular organs and the central actions of angiotensin. Neuroendocrinology 32:248–256

Simpson JB, Routtenberg A (1975) Subfornical organ lesions reduce intravenous angiotensin-induced drinking. Brain Res 88:154–161

Sladek CD, Joynt RJ (1979) Cholinergic involvement in osmotic control of vasopressin release by the organ-cultured rat hypothalamo-neurohypophyseal system. Endocrinology 105:367

Sladek CD, Knigge KM (1977) Cholinericg stimulation of vasopressin release from the rat hypothalamo-neurohypophyseal system in organ culture. Endocrinology 101:411–420

Sofroniew MV, Glasmann W (1981) Golgi-like immunoperoxidase staining of hypothalamic magnocellular neurons that contain vasopressin, oxytocin or neurophysin in the rat. Neuroscience 6:619–643

Sofroniew MV, Weindl A (1978) Extrahypothalamic neurophysin-containing perkarya, fiber pathways and fiber clusters in the rat brain. Endocrinology 102:334–337

Stamler JF, Phillips MI (1981) Attenuation of the central hypertonic NaCl pressor response by angiotensin II inhibition. Brain Res 213:427–431

Stefanini E, Devoto P, Marachisio AM, Vernaleone F, Collu R (1980) [^{3}H] Spiroperidol binding to a putative dopaminergic receptor in rat pituitary gland. Life Sci 26:583–587

Summerlee AJS (1981) Extracellular recordings from oxytocin neurones during the expulsive phase of birth in unanaesthetized rats. J Physiol (Lond) 321:1–9

Summerlee AJS (1983) Hypothalamic neurone activity: hormone release and behaviour in freely moving rats. Q J Exp Physiol 68 (in press)

Summerlee AJS, Lincoln DW (1981) Electrophysiological recordings from oxytocinergic neurones during suckling in the unanaesthetized lactating rat. J Endocrinol 90:255–265

Summerlee AJS, Paisley AC (1982) The effect of behavioural arousal on the activity of hypothalamic neurons in unanaesthetized, freely moving rats and rabbits. Proc R Soc Lond [Biol] 214:263–272

Summy-Long JY, Keil LC, Deen K, Severs WB (1981 a) Opiate regulation of angiotensin-induced drinking and vasopressin release. J Pharmacol Exp Ther 217:630–637

Summy-Long JY, Rosella LM, Keil LC (1981 b) Effects of centrally administered endogenous opioid peptides on drinking behaviour, increased plasma vasopressin concentration and pressor response to hypertonic sodium chloride. Brain Res 221:343–357

Summy-Long JY, Severs WB (1979) Macromolecular changes in the subfornical organ area after dehydration and renin. Am J Physiol 237:R26–38

Swanson LW, Kuypers HGJM (1980) The paraventricular nucleus of the hypothalamus: cytoarchitectonic subdivisions and organization of projections to the pituitary, dorsal vagal complex and spinal cord as demonstrated by retrograde fluorescence double-labeling methods. J Comp Neurol 194:555–570

Swanson LW, Sawchenko PE, Rivier J, Vale WW (1983) Organization of ovine corticotropin-releasing factor immunoreactive cells and fibers in the rat brain: an immuno-histochemical study. Neuroendocrinology 36:165–186

Theodosis DT (1982) Secretion-related accumulation of horseradish peroxidase in magnocellular cell bodies of the rat supraoptic nucleus. Brain Res 233:3–16

Theodosis DT, Poulain DA, Vincent JD (1981) Possible morphological bases for synchronisation of neuronal firing in the rat supraoptic nucleus during lactation. Neuroscience 6:919–929

Tribollet E, Clarke G, Dreifuss JJ, Lincoln DW (1978) The role of central adrenergic receptors in the reflex release of oxytocin. Brain Res 142:69–84

Tweedle CD, Hatton GI (1977) Ultrastructural changes in rat hypothalamic neurosecretory cells and their associated glia during minimal dehydration and rehydration. Cell Tissue Res 181:59–72

Unger T, Rascher W, Schuster C, Pavlovitch R, Schomig A, Dietz R, Ganten D (1981) Central blood pressure effects of substance P and antiotensin II: role of the sympathetic nervous system and vasopressin. Eur J Pharmacol 71:33–42

Urban IJA (1981) Intraseptal administration of vasopressin and oxytocin affects hippocampal electroencephalogram in rats. Exp Neurol 74:131–147

Vale MR, Hope DB (1982) Cyclic nucleotides and the release of vasopressin from the rat posterior pituitary gland. J Neurochem 39:569–573
Vandesande F, Dierickx K (1975) Identification of the vasopressin-producing and of the oxytocin-producing neurons in the hypothalamic magnocellular neurosecretory system of the rat. Cell Tissue Res 164:153–162
Vandesande F, Dierickx K, De May J (1977) The origin of the vasopressinergic and oxytocinergic fibres of the external region of the median eminence of the rat hypophysis. Cell Tissue Res 180:443–452
Van Leeuwen F (1980) Immunocytochemical specificity for peptides with special reference to arginine-vasopressin and oxytocin. J Histochem Cytochem 28:479–482
Van Ree JM, Bohus B, Versteeg DHG, De Wied D (1978) Neurohypophysial principles and memory processes. Biochem Pharmacol 27:1793–1800
Verney EB (1947) The antidiuretic hormone and the factors which determine its release. Proc R Soc Lond [Biol] 135–25–105
Vincent SR, Hokfelt T, Wu J-Y (1982) GABA neuron systems in hypothalamus and the pituitary gland. Neuroendocrinology 34:117–125
Vizi ES, Volbekas V (1980) Inhibition by dopamine of oxytocin release from isolated posterior lobe of the hypophysis of the rat; disinhibitory effect of β-endorphin/enkephalin. Neuroendocrinology 31:46–52
Voloschin LM, Tramezzani JH (1979) Milk-ejection reflex linked to slow-wave sleep in nursing rats. Endocrinology 105:1202–1207
Wakerley JB, Lincoln DW (1971) Phasic discharge of antidromically identified units in the paraventricular nucleus of the hypothalamus. Brain Res 25:192–194
Wakerley JB, Lincoln DW (1973) The milk-ejection reflex of the rat: a 20- to 40-fold acceleration in the firing of paraventricular neurones during oxytocin release. J Endocrinol 57:477–493
Wakerley JB, Noble R (1982) Electrophysiological behaviour of putative vasopressin neurones in intact brains and hypothalamic slices. In: Baertschi AJ, Dreifuss JJ (eds) Neuroendocrinology of vasopressin, corticoliberin and opiomelanocortins. Academic, London
Wakerley JB, Dyball REJ, Lincoln DW (1973) Milk ejection in the rat: the result of a selective release of oxytocin. J Endocrinol 57:557–558
Wakerly JB, Poulain DA, Dyball REJ, Cross BA (1975) Activity of phasic neurosecretory cells during haemorrhage. Nature 258:82–84
Wakerley JB, Poulain DA, Brown D (1978) Comparison of firing patterns in oxytocin- and vasopressin-releasing neurones during progressive dehydration. Brain Res 148:425–440
Wakerley JB, Noble R, Clarke G (1983 a) In vitro studies of the control of phasic discharge in neurosecretory cells of the supraoptiv nucleus. In: Cross BA, Leng G (eds) The neurohypophysis – structure, function and control. Elsevier, Amsterdam
Wakerley BJ, Noble R, Clarke G (1983 b) Effects of morphine and D-Ala, D-Leu enkephalin on the electrical activity of supraoptic neurosecretory cells in vitro. Neuroscience 10:73–81
Wang BC, Share L, Crofton JT, Kimura T (1982) Effect of intravenous and intracerebroventricular infusion of hypertonic solutions on plasma and cerebrospinal fluid vasopressin concentrations. Neuroendocrinology 34:215–221
Watson SJ, Akil H, Fischli W, Goldstin A, Zimmerman E, Nilaver G, van Wimersma Greidanus TB (1982) Dynorphin and vasopressin: common localization in magnocellular neurons. Science 216:85–87
Weindl A, Sofroniew MV (1981) Relation of neuropeptides to mammalian circumventricular organs. Adv Biochem Psychopharmacol 28:303–320
Yagi K, Azuma T, Matsuda K (1966) Neurosecretory cell: capable of conducting impulse in rats. Science 154:778–779
Zerihun L, Harris M (1981) Electrophysiological identification of neurones of paraventricular nucleus sending axons to both the neurohypophysis and the medulla in the rat. Neurosci Lett 23:157–160

Volume Regulation of Antidiuretic Hormone Secretion*

M. J. McKINLEY[1]

Contents

1 Introduction

In simplest terms, the volume of fluid in the body of the mammalian organism depends on the balance between the amount of water ingested or metabolically produced and the amount lost from the body by respiratory, excretory, or cooling processes. With regard to loss of body water, there is little doubt that the major homeostatic regulation is effected by the action of the antidiuretic hormone (ADH) on the kidneys. This neurohypophysial hormone, which in most mam-

* Work supported by grants from the National Health and Medical Research Council of Australia

1 Howard Florey Institute of Experimental Physiology and Medicine, University of Melbourne, Parkville 3052, Victoria, Australia

mals is arginine vasopressin (AVP), reduces urinary water loss by causing renal reabsorption of free water from the glomerular filtrate. If one needs convincing of the importance of ADH for maintaining body fluid volume, one need go no further than observing the spectacular output of urine by an individual or animal afflicted with hypothalamic diabetes insipidus and the subsequent decline of urine flow when ADH is administered. Indeed, the fascination with which earlier physicians and investigators viewed this disorder was almost certainly the major impetus leading to the genesis of this field of physiology in the early years of this century (Frank 1910; Farini 1913; von den Velden 1913).

Obviously ADH action alone is insufficient to maintain the fluid volume of the body. Adequate water intake is essential for maintenance of body fluids, otherwise loss of water from the lungs, skin, and kidneys will inexorably lead to severe dehydration and death. In addition, if the composition (particularly the osmolarity and sodium concentration) of body fluids is maintained within narrow limits, body sodium balance will be an important factor influencing the volume of these fluids. Consequently, regulation of the volume and distribution of body water is integrally linked with osmoregulatory mechanisms. Such regulation involves (a) behavioral processes of water, food, and sodium intake, (b) a plethora of hemodynamic, endocrine, and neural mechanisms regulating renal excretion of solutes, and (c) the fine control of renal free water excretion by ADH. It is beyond the scope of this article to deal with all these factors, and the following paragraphs will be devoted to a consideration of recent contributions to our knowledge of the physiological relationship between altered body fluid volume and secretion of ADH. Comprehensive information on aspects of the molecular synthesis, anatomical distribution and cellular mechanism of release of ADH is provided in the works of Brownstein et al. (1980), Sladek et al. (1980), and Land et al. (1982).

Probably as a result of the historical development of knowledge in this field, commentaries on the physiological release of ADH have often categorized the factors involved as being of an osmotic or nonosmotic nature (Robertson 1977; Forsling 1977; Schrier et al. 1979). This convenient classification will be utilized here, but it seems appropriate in the context of this review to relate such a classification to the associated alterations in fluid volume of the various body compartments. Body water may be broadly divided into two major compartments, the intracellular phase (comprising two-thirds of the total) and the extracellular phase (comprising the interstitial and vascular water). Andersson (1978) has drawn up a convenient scheme to describe the way changes in the volume of the intracellular and extracellular compartments relate to thirst and ADH secretion. As Andersson points out, thirst and hypersecretion of ADH are most effectively stimulated by situations in which cellular dehydration occurs. This is not surprising in view of the fact that a majority of the body's water is contained within cells, and increased ADH levels should minimize further water losses, whilst thirst with subsequent water intake should replenish both cellular and extracellular water. The common features associated with cellular dehydration are (a) reduced cell volume, (b) increased sodium concentration of plasma, and (c) hyperosmolarity of body fluids. Regardless of whether it is the altered cell volume (Verney 1947) or the changed sodium concentration (Andersson 1978) which triggers the detector mechanism initiating ADH release, regulation of cellular volume is inseparably linked with osmotic factors and the ambient sodium concentration. In contrast to the ADH secretion which arises from hypertonicity and is associated with cellular dehydration, reduction in the extracellular fluid (ECF) volume without decreased intracellular fluid (ICF) volume may also lead to increased ADH secretion, and conversely, isotonic expansion of the ECF volume may result in reduced levels of ADH. This altered ECF volume may be one of a number of "nonosmotic" factors influencing ADH secretion. Other regulators of ADH release which may be viewed as having homeostatic significance include body temperature and behavioral factors associated with water intake.

2 Intracellular Fluid Volume and ADH Secretion

2.1 Osmotic Factors Regulating ADH Secretion: Theories on the Sensor Mechanisms

The concept that tonicity of the ECF is a major determinant of the rate of ADH secretion has been unequivocally accepted for many years. Many workers have observed that after an initial threshold level of plasma osmolality is reached, there is a close relationship between the osmolality of plasma and its ADH concentration (Dunn et al. 1973; Robertson and Athar 1976; Bayliss and Robertson 1980; Quillen et al. 1983). This relationship appears to be linear over the greater part of the physiological range of plasma osmolality for all species studied, with the threshold and gradient varying between species (Fig. 1). In order to find the origins of this concept we must turn to the work of Verney, whose major contribution to this field was concerned with the nature and location of the sensors detecting changes in the effective osmolarity of ECF. During the war years and in the immediate postwar period, he carried out a series of studies on conscious dogs which laid the foundations of much of our present knowledge of this subject. Very briefly, Verney found that by selectively elevating the effective osmotic pressure of blood perfusing the field of distribution of the internal carotid artery, he could effect a rapid reduction in urine output of dogs undergoing a water diuresis. He attributed this to the release of an antidiuretic substance from the posterior pituitary, postulated that the detection of blood hypertonicity was the function of cells in the diencephalon, and coined the term osmoreceptors to describe these cells (Verney 1947). In subsequent studies with Jewell, he was able to further clarify, although not completely pinpoint, the location of these cerebral osmoreceptors to the anterior hypothalamic/preoptic area of the brain (Jewell and Verney 1957). His work was such a tour de force that for the following 25 years little further investigation of this topic was made. If that period was notable for a lack of further debate on the osmoregulation of vasopressin, the same cannot be said of the decade just past. In fact, the osmoreceptors have been described as "enigmatic" (Forsling 1977) and compared to "the Holy Ghost in that they are often talked about but nobody is sure of having seen them" (Andersson 1978), and their existence has been virtually denied in that "the osmoreceptor hypothesis of Verney is incompatible with a number of apparently valid results" (Bie 1980). The

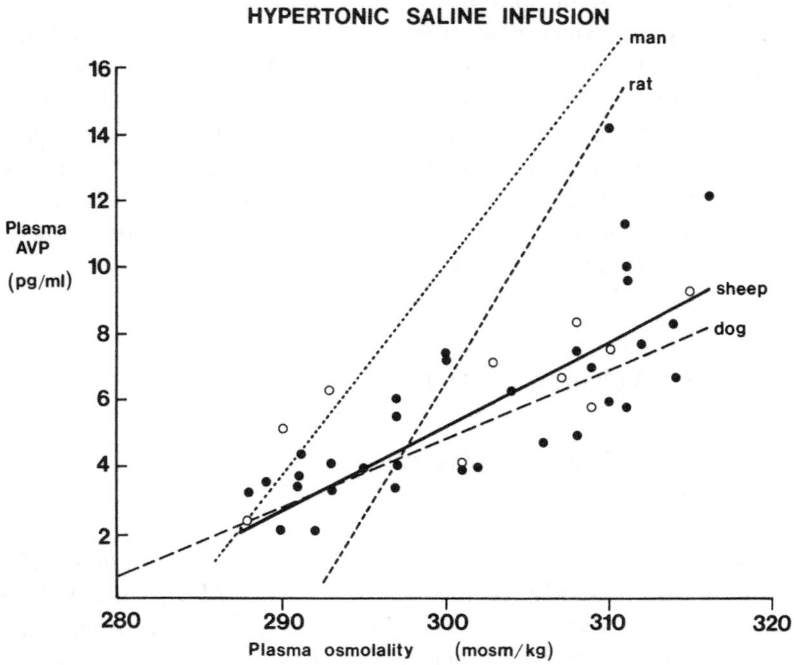

Fig. 1. The relationship between plasma [AVP] and plasma osmolality (pOsm) when pOsm was progressively increased by infusion of 4 *M* NaCl at 1.3 ml/min for 40 min into the jugular vein of five unanesthetized sheep (*closed circles*). Blood samples were taken at 10-min intervals from a cannula placed in the carotid artery and plasma [AVP] was measured by radioimmunoassay. The linear regression is given by the equation plasma [AVP] = 0.26 (pOsm–280)] $r = 0.81$, $p < 0.01$, and is shown by the *solid line*. Plasma osmolality was also increased in three sheep by intravenous infusion of 1 *M* mannitol in normal saline and the results are shown by the *open circles*. These points were not used in the linear regression analysis. For comparison (*interrupted lines*) the regression lines obtained by other workers who induced hyperosmolarity in various ways in other species have been redrawn from their papers; Dunn et al. (1973) for the rat, Bayliss and Robertson (1980) for man, and Quillen and Cowley (1983) for the dog

three major bones of contention are (a) whether cerebral osmoreceptors actually exist (Bie 1976, 1980); (b) whether osmoreceptors are really sensor cells which respond specifically to changes in the [Na] of the cerebrospinal fluid (CSF) (Andersson 1977; McKinley et al. 1978), and (c) whether the neurosecretory cells of the supraoptic nucleus (SON) which synthesize ADH are actually the site of osmoreceptors (Jewell and Verney 1957; Leng 1980; Leng et al. 1982). These three issues will now be examined.

2.2 Cerebral Osmoreceptors: Fact or Fiction?

The pioneering work of Verney on this topic has already been noted. However, in 1980 Bie launched a direct assault on Verney's evidence for cerebral osmoreceptors. Because Verney's work is the cornerstone of much of our present knowledge of ADH secretion, this section will examine and offer rebuttals to the objections raised by Bie. In his detailed review (1980), Bie critically examined the results of many investigators and concluded that the evidence so far produced was inadequate to sustain the osmoreceptor hypothesis. No doubt these views were partly influenced by his own results obtained in hydrated dogs which had been anesthetized with chloralose (Bie 1976). In these experiments he found that bilateral infusion of hypertonic NaCl solution into the carotid arteries was no more effective than an equivalent intravenous infusion in reducing renal free water clearance, casting doubt on the view that hypothalamic osmoreceptors play an important role in regulating plasma ADH levels. Later studies (Bie et al. 1982) in which direct measurements of plasma ADH were made in chloralosed dogs also supported such contention. In the light of these findings, it is interesting to examine the objections to Verney's findings and hypothesis which were raised in review by Bie (1980). While a number of valid points of criticism were made by Bie, there were four major objections.

Firstly, after graphing data from Verney's other papers on the rate of excretion of a water load by dogs (Klisiecki et al. 1933; O'Connor and Verney 1942), Bie claimed that the rate of reduction in urine flow during an intracarotid infusion of hypertonic sucrose solution was the same as if no infusion had been made. This, however, is an oversimplification of the question; in point of fact, the more appropriate control experiment to the intracarotid infusion was the intravenous infusion of hypertonic sucrose, which Verney did report. He showed such an infusion to have less antidiuretic effect than the intracarotid infusion of sucrose. Because systemically infused sucrose may induce an osmotic diuresis (Verney showed reducing sugar in the urine), one would expect more prolonged diuresis in animals after water loading and systemic infusion with sucrose than after water loading alone. Thus intracarotid infusion of sucrose did have an antidiuretic effect appropriate to the water load and the systemic load of sucrose. This was further confirmed by Verney's observation that it was necessary to infuse a particular amount of posterior pituitary extract in combination with intravenous sucrose to reproduce the antidiuretic effect of the infusion of hyperosmolar sucrose into the carotid artery.

Bie's second major objection to Verney's conclusion was based on the fact that some injections of hypertonic solutions appeared to cause discomfort and emotional disturbance to dogs, phenomena which in themselves might induce ADH release (O'Connor and Verney 1942). Verney did indeed report apparent discomfort to dogs in some experiments, but these were experiments where rapid increase by over 50% in carotid blood osmolarity was induced and rapid antidiuresis was reported to occur. No mention is made of adverse effects of slow intracarotid infusions of hypertonic NaCl or sucrose which raised carotid blood osmolarity by only 2%–5% and which were antidiuretic, and it was these latter experiments which were crucial in establishing a physiological role for the cerebral osmoreceptors. Verney was careful to eliminate the possibility that discomfort to the dog was caused by these infusions, and in view of the meticulously detailed reporting which is evident throughout Verney's publications, it would seem unlikely that any such adverse effects occurred.

Another challenge to Verney's findings was made on the basis that the conclusions were derived from results obtained in only two animals. While this may be the case in the Croonian Lecture of 1947 (and does not necessarily invalidate the results), further verification of the major conclusion is provided in the results of Jewell and Verney (1957). In this latter paper, it was shown that the antidiuresis caused by intracarotid infusion of hypertonic NaCl was diminished when various branches of the internal carotid artery were ligated. In this set of experiments, a number of rapid antidiuretic responses to intracarotid infusion of hypertonic NaCl was reported in 13 dogs, confirming the earlier studies. This paper in particular was taken to task by Bie (1980). Although strictly speaking Bie was correct in stating that in "9 of 12 dogs the expected effect of intracarotid infusions of hyperosmolar NaCl was frequently absent" and "the results from intravenous infusions of hyperosmolar infusions appear equally unpredictable," he appears to have overlooked two crucial considerations in analyzing this data. Firstly, in the initial section of their publication, Jewell and Verney showed by tracing the field of distribution of the carotid blood flow that there was "more or less asymmetry of distribution of carotid blood of the two sides between anterior cerebral structures." They went on to say that "one carotid normally dominated and supplied not only the structures of the ipsilateral medial sides but the medial region of the contralateral side as well." In view of this evidence, it might be expected that the antidiuretic effect of intracarotid administration of hyperosmolar solutions would be more marked on infusion into one carotid than into the other, particularly if the osmoreceptors were somewhat medially located, a suggestion which has recently been made (McKinley et al. 1978; Thrasher et al. 1982a). In fact it can be seen that this was the case. In all 13 dogs that Jewell and Verney examined, it is clear that hypertonic NaCl caused a rapid antidiuresis when infused into one of the two carotids. While there were some antidiuretic responses seen with intravenous infusions of hypertonic solutions, these responses were always slower than those obtained with the effective intracarotid infusion. Therefore these results are entirely in agreement with those which would be predicted on the basis of the existence of a cerebral osmoreceptor.

The fourth major criticism which Bie (1980) made of Verney's osmoreceptor hypothesis was that it had not been verified by others. Studies in water-loaded goats and sheep which purported to reproduce the observations of Verney (1947) regarding stronger antidiuretic responses with intracarotid infusions of hypertonic solution than with intravenous administration also came under close scrutiny from Bie. It is clear from several papers that infusion of hypertonic NaCl or fructose into the carotid artery of water-loaded goats causes a rapid reduction in renal free water clearance (Eriksson et al. 1971; Olsson 1972, 1973; Olsson and Kolmodin 1974). There was evidence that such infusions did not cause any discomfort to the animals, that the effect was absent in animals having a deficiency of ADH due to experimentally induced diabetes insipidus, and that the increase in the osmolality of the blood was of the order of 13 mosm/kg, well within the range of osmolality change observed in dehydrated goats (Eriksson et al. 1971). While not all of these papers gave specific evidence of the effects of intrajugular administration of similar hypertonic solutions, detailed results were shown in one report (Olsson 1973). Despite Bie's assertions that in "these experiments the intravenous administration of hyperosmolar saline resulted in an antidiuretic response of the same order of magnitude as that produced by the intracarotid infusion" this is clearly not so. The rate of reduction of renal free water clearance with intracarotid infusions of hypertonic saline was approximately double that which occurred with intravenous infusions. For instance, renal free water clearance fell on average by 1.5, 4.5, and 6.5 ml/min respectively at 10, 20, and 30 min after the commencement of intracarotid infusion of 1 *M* NaCl. In contrast, this parameter fell by <0.5, 2, and 4 ml/min during the corresponding periods with intrajugular infusion of 1 *M* NaCl at a similar rate. In addition, the renal free water clearance remained negative for over 50 min after the end of intracarotid infusion of hypertonic saline, but was always positive following the intrajugular infusion. Bie's claim that increased sodium excretion could have accounted for the greater reduction in free water clearance with the intracarotid infusions is also doubtful. In fact, other workers have shown that the ability of the kidney to concentrate urine in response to ADH is diminished during times of increased solute excretion (Orloff et al. 1958; Gottschalk 1964). Despite the fact that Olsson and colleagues (Eriksson et al. 1971; Olsson 1973; Olsson and Kolmodin 1974) may have put a dissenting interpretation on these results in goats, in that the sensors responded to sodium concentration rather than osmolality, they clearly replicate Verney's findings and are consistent with a cerebral sensor detecting systemic hypertonicity, whether it be a sodium sensor or an osmoreceptor.

With regard to our experiments in sheep (Blaine et al. 1975; McKinley et al. 1978), it was not our initial purpose to test specifically the validity of the hypothesis that cerebral osmoreceptors regulate ADH secretion. Nevertheless, these experiments were made in water-loaded sheep, which allowed comparison of the antidiuretic effects of intracarotid and intrajugular administration of hypertonic solutions. The results obtained (Blaine et al. 1975) were consistent with those of Verney (1947) and Olsson (1973). That is, infusion of hypertonic solutions of NaCl, sucrose, or fructose exhibited greater antidiuretic effects when infused into a carotid artery than into the jugular vein of conscious, water-loaded sheep. Several criticisms were levelled by Bie (1980) at these findings and some explanation may help to allay his misgivings. Firstly, Bie considered that occlusion of one carotid artery during infusion into the other was less than desirable. In explanation, this was done to provide bilateral distribution of the infusates into the carotid artery (Baldwin and Bell 1963). Such a procedure does not induce secretion of ADH in sheep (McKinley et al. 1978) as it does in other species, because in the sheep there is only a monentary drop in arterial blood pressure above the occlusion due to an effective occipitovetebral anastomosis between the fields of distribution of the two carotid arteries. Additionally, the possible problem of asymmetric distribution of carotid blood between the two sides of the brain, as encountered by Jewell and Verney (1957), is overcome, and in any case the same procedure of occluding one carotid was also carried out during the comparative infusion into the jugular vein. The second criticism of this work concerned possible adverse effects of infusions of sucrose and fructose. In the grouped experiments reported, particular care was taken to exclude any trials in which adverse effects were observed. Blood pressure changed in only one of the 12 trials included in this grouped data, a fact which strongly supports the view that pain and discomfort were not influencing the secretion of ADH in these experiments. When one considers the relatively slow antidiuretic response caused by intracarotid infusion of 2 *M* urea in comparison to infusion of 1 *M* NaCl or 2 *M* fructose (McKinley et al. 1978), confirmation of a further facet of Verney's original data is afforded, in that the effective osmotic pressure rather than the absolute osmolarity of body fluids is the determinant of ADH secretion. A proviso which we added, however, was that the osmoreceptors were probably located in a brain region lacking the blood-brain barrier, and this notion will be further discussed in a later section 2.3. An unexpected observation in these studies in sheep (McKinley et al. 1978) was that continued intracarotid administration of hyperosmolar urea did eventually elicit a strong antidiuretic response, although slower than with infusions of NaCl or fructose. Although Bie claimed (1980) that we considered that the effects of urea infusions could be due to a deleterious action of this agent, this was an erroneous interpretation and in point of fact evidence was presented to show this was not the case. Such data could be consistent with an additional osmoreceptor or sodium sensor situated behind the blood-brain barrier (McKinley et al. 1978). The question is also raised of whether different sensor mechanisms may be operative in the longer term to detect systemic hypertonicity than those involved in detecting short-term (20-min) changes in tonicity.

Studies in conscious rhesus monkeys have also been reported (Swaminathan 1980). Once again, it was found that infusion of hyperosmolar solutions of sodium chloride, sodium acetate, sucrose, or mannitol caused greater antidiuretic responses when infused into the carotid artery than into a limb vein. Infusions of NaCl caused the greatest effects. While this data is consistent with the concept of a cerebral sensor regulating ADH secretion in primates, some reservations must remain, as this investigator reported all infusions of the test solutions to cause mild arousal and transient uneasiness, which may have influenced the result.

All the studies discussed so far utilized an indirect indication of plasma ADH levels, namely urine flow rate, urine osmolality, or renal free water clearance. Recently, the plasma ADH levels measured in experiments in which selective increases in cerebral blood osmolality were contrived by means of intracarotid infusions in conscious dogs have been reported (Dietz et al. 1982; Wade et al. 1982a, b). The findings of Dietz et al. are not incompatible with the concept of a cerebral osmoreceptor. The more recent findings of Wade et al., compared the effects of short-term (6-min) bilateral infusions of hypertonic NaCl into the carotid arteries or saphenous veins of conscious dogs and observed increased plasma AVP levels only with intracarotid infusions. These results appear to be definitely in accord with the osmoreceptor hypothesis. In view of the fact that Bie, principal inquisitor of Verney's work, was a coauthor of both of these papers, much of the controversy discussed above now seems to be resolved, with the disparity between his earlier results and these recent findings being attributed to an undefined influence of chloralose anesthesia on the function of the cerebral osmoreceptors.

2.3 Cerebral Osmoreceptors or Sodium Sensors?

During the last decade a number of publications, principally from laboratories in Sweden and Finland, have suggested that the cerebral osmoreceptors are really sensors which detect changes in the [Na] of the CSF. The chief protagonists of this view – Andersson, Olsson, and associates in Stockholm and Uppsala – propose that these sensors are located in the anterior wall of the third cerebral ventricle and regulate ADH secretion, thirst, and renal output of sodium (for reviews see Andersson 1977, 1978).

What is the main evidence leading these workers to such a conclusion? Initially, they were impressed by the inability of injections of hypertonic solutions of nonelectrolytes such as sucrose, glucose, and fructose into the third cerebral ventricle to elicit an antidiuresis in water-loaded goats (Andersson et al. 1967; Olsson 1969; Eriksson et al. 1971). In contrast, intraventricular administration of hypertonic NaCl was effective in stimulating ADH secretion and an antidiuresis (Andersson et al. 1967; Olsson 1969; Fyhrquist et al. 1979; McKinley et al. 1980). Subsequently, they have shown that the antidiuretic responses to intracarotid infusions of hypertonic NaCl could be blocked by lowering the [Na] of CSF by means of simultaneous intracerebroventricular (i.c.v.) infusion of hypertonic saccharide solutions (Olsson 1972; Olsson 1973). Additionally, it has been found that i.c.v. infusion of isotonic or hypertonic solutions of saccharides which lowered CSF [Na], but which would either increase or cause no change in osmolality of CSF, caused inhibition of ADH secretion resulting in subsequent water diuresis (Eriksson 1974; Fyhrquist et al. 1979). Because infusion into the anterior chamber of the third cerebral ventricle was the most effective route of administration of hypertonic saline for eliciting an antidiuresis, and because ablation of the anterior wall of the third ventricle abolished this response, Andersson and co-workers proposed that this region was the site of sodium sensors (Andersson et al. 1975). This group of investigators was also able to show that i.c.v. administration of agents which block transmembrane sodium transport, such as glycerol (Olsson et al. 1976), deuterium (Leksell et al. 1976), and ethacrynic acid (Rundgren et al. 1979), inhibited vasopressin secretion in the goat. They suggested that alteration in Na-K ATPase activity may have been part of the mechanism by which a sodium sensor operates, although it could be argued that inhibition of sodium extrusion from cells would lead to cell swelling, and the observed inhibition of ADH secretion by inhibitors of sodium transport could also be consistent with the operation of an osmoreceptor.

Following these extensive studies made in goats, investigations into the feasibility of cerebral sodium sensors were made in sheep (McKinley et al. 1978), dogs (Thrasher et al. 1980a), and monkeys (Swaminthan 1980), and results have not been entirely concordant. In sheep it was found that an antidiuretic response to i.c.v. infusion of hypertonic sucrose or fructose solution could be obtained provided that the solutions also contained basal CSF concentrations of Na^+ and other cations. Such evidence was construed as evidence in favor of osmoreceptors in sheep (McKinley et al. 1978). Nevertheless, the antidiuresis in response to intrathecally administered saccharides was of sufficiently smaller magnitude in comparison with that observed after an equivalent i.c.v. infusion of hypertonic NaCl for a specific effect of the [NaCl] over and above osmotic consideration to be indicated (McKinley et al. 1978). Subsequent studies in goats have confirmed and extended these results in that the effects at the kidney were found to be due to higher plasma ADH levels with i.c.v. infusions of hypertonic NaCl (McKinley et al. 1980). Lowering of the [Na] but not the osmolality of CSF in water-replete or dehydrated sheep also resulted in reduced ADH levels in blood and a resultant water diuresis (Leksell et al. 1981). Results obtained in rhesus monkeys also implicated involvement of a cerebral sodium sensor in ADH secretion in primates (Swaminthan 1980). In contrast to these studies were the results of Thrasher et al. (1980a), who found that i.c.v. administration of hypertonic sodium solution was no more effective in stimulating ADH release in dogs than an equivalent infusion of hypertonic sucrose into the CSF. Thus in the dog, no evidence for a cerebral sodium sensor involved in ADH secretion was forthcoming. Utilizing a different approach, the feasibility of cerebral sodium sensors was further tested in sheep and dogs by measuring the changes in CSF [Na] which occurred as a consequence of systemic administration of hyperosmolar solutions of NaCl, sucrose, or urea (McKinley et al. 1978; Thrasher et al. 1980b). The common finding of both groups was that in the short term, ADH secretion did not correlate with changes in the CSF [Na], and therefore a sensor of CSF [Na] could not explain the results. The data seemed more compatible with the concept that ADH secretion was under the control of a cerebral osmoreceptor located outside the blood-brain barrier. In the longer term (30–45 min), however, there was a difference in the antidiuretic response to systemic infusion of hypertonic urea between sheep and dogs. Although no antidiuresis or ADH release was initially seen in either species, systemic infusion of hyperosmolar urea eventually resulted in antidiuresis in the sheep (McKinley et al. 1978). Because increased CSF [Na] was another consequence of the urea infusion, the late antidiuresis observed with urea infusion could be due to stimulation of either a sodium receptor or an osmoreceptor located behind the blood-brain or blood-liquor barrier in sheep. Whatever the mechanism of this late stimulation of ADH secretion by urea infusion in sheep might be, it is clear that the major evidence for cerebral sodium sensors is derived from experiments in sheep and goats, and it is possible that the intracerebral [Na] is of more significance for the regulation of ADH secretion and fluid balance in ruminant species. Certainly these ruminants appear to be superbly adapted to survival in arid regions of the earth (McFarlane and Howard 1972), and certain antidiuretic mechanisms may be more highly developed in sheep and goats than in some other species. From a teleological point of view, whether the sensors under discussion are responding to soduim concentration, osmotic pressure, or both, in most cases the really effective changes in osmolality of body fluids will result from changes in [Na]. Another observation which gives cause for thought is that with prolonged dehydration, brain cells tend to revert to normal size after the initial cell shrinkage (Arieff and Guisado 1976). If this happened in the case of the cerebral osmoreceptors (whose operation was thought to depend on osmotically induced alterations in cell volume), their functional usefulness in maintaining ADH secretion would clearly be compromised. It could be that a sensor based on the intracerebral fluid concentration of sodium may be more theoretically attractive in that such functionally effective accommodation would be less likely with a sodium sensor because the raised [Na] is maintained throughout dehydration (Leksell et al. 1981). Because the experiments with systemic infusions of hypertonic solutions alluded to above involved observations lasting no longer than 45 min, studies into the osmotic regulation of ADH secretion over longer periods are probably warranted. However, it seems reasonable to conclude that while there is good evidence in most species for the existence of cerebral osmoreceptors (and in ruminants additional sodium sensors), exact characterization of their responsiveness will have to await precise location in the brain of these detector elements.

2.4 Where Are the Cerebral Osmoreceptors Located?

If cerebral osmoreceptors exist, the next step in analysis concerns their precise location. As mentioned earlier, Jewell and Verney (1957) embarked on the first systematic search for these sensors and were able to eliminate many brain regions, as well as the posterior pituitary gland, carotid sinus, and carotid body. The regions which could not be eliminated, and therefore within which the osmoreceptors were probably located, were the anterior hypothalamic and medial preoptic regions. The method used by Jewell and Verney involved testing the antidiuretic responses in conscious dogs to intracarotid infusions of hypertonic NaCl before and after hemispherectomy or before and after various branches of the internal carotid artery had been intradurally ligated in order to redistribute the blood carrying the hyperosmolar stimulus to various brain regions. The distribution of blood flow from each carotid artery was then determined by histological examination after different coloured suspensions had been administered into each carotid just before the dogs were killed. While Jewell and Verney devoted special attention to the SON and concluded that their evidence was "not merely consistent, but strongly supportive of the view that the receptors are in or in the close region of this nucleus," they went on to mention other regions, stating that "the claims of some of them, the medial preoptic area, the suprachiasmatic nucleus, the nucleus supraopticus diffusus, the anterior hypothalamic area, the periventricular system and the ventromedial nucleus must for the present be regarded as not inferior to those of the supraoptic nucleus itself." In fact, there was evidence that the SON could not have been an exclusive site of the osmoreceptors; in one of the dogs studied by Jewell and Verney this region had been perfused with hyperosmolar solution and no antidiuretic response obtained. Because cystic damage to the paraventricular nucleus (PVN) was observed in this dog, it was proposed that an intact PVN was essential for the operation of the osmoreceptors, but an alternative interpretation is that in fact the SON was not the site of the osmoreceptors. Consistent with the findings of Jewell and Verney (1957) were the later studies of Woods et al. (1966), who investigated ADH release in decerebrated dogs and found that hypothalamic-preoptic islets of tissue were sufficient to maintain osmotic control of ADH release. Somewhat analogous to this result is the observation that hypothalamoneurohypophysial explants from the rat will respond to hypertonic stimuli with increased ADH release into their bathing media (Sladek 1980; Ishikawa et al. 1980).

These studies suggest that some tissue within the general area of the anterior hypothalamic-preoptic brain region is crucial for the osmotic regulation of ADH and provide a background to investigations aimed at locating the osmoreceptors more precisely. Perhaps the question which most have attempted to answer is whether the osmoreceptors are also the neurosecretory cells of the SON. A number of investigators have made electrophysiological recordings from cells in the SON and found changes in the firing rate during intracarotid infusions of hyperosmolar solutions (Koizuma et al. 1964; Vincent and Hayward 1970; Vincent et al. 1972a). However, cells in other brain regions have also been found to alter their electrical activity with alterations in osmolarity of carotid artery blood, e.g., the perinuclear zone of the SON (Hayward and Jennings 1973), the PVN (Brooks et al. 1966), the lateral hypothalamic region (Weiss and Almli 1975), and the lateral preoptic region (Malmo and Malmo 1979), to mention but a few. Unfortunately, this methodology does not allow us to determine whether activity of the cells changed in response to a direct action of the osmolarity change or to inputs received from other osmotically sensitive units elsewhere in the brain. More recently, hyperosmolar stimuli have been directly applied by microtap to cells of the SON, and these neurons, identified by antidromic stimulation from the pituitary stalk, responded to hypertonic NaCl with altered electrical activity (Leng 1980). Leng also estimated that the changes in tonicity induced in the vicinity of the neurons under study were probably within physiological bounds. Although such findings were not initially observed by others with in vitro recording from hypothalamic slices (Haller and Wakerley 1980), a later report indicates that under certain conditions of continuous exposure to glutamate, SON neurons will respond to increased NaCl (but not mannitol) concentration of the bathing medium, and possibly function as a sodium sensor (Noble and Wakerley 1982). Some workers have suggested that even though ADH-secreting neurons may be identified as the phasically firing cells identified by antidromic stimulation from the pituitary, there may be other ADH-secreting cells which do not meet these criteria (Yagi et al. 1980). Further research to clarify the fraction of ADH-synthesizing cells which respond to applied hypertonicity will be of value. In addition, the apparent disagreement as to whether the neurosecretory cells of the SON respond to change in osmotic pressure or NaCl concentration is worthy of resolution.

Another technique which has been employed is the application of hypertonic solutions by microinjection into a range of brain loci. Using this strategy in rats, Peck and Blass (1975) found the most sensitive cerebral regions to directly applied altered tonicity were the anterior hypothalamic and preoptic regions and the bed nucleus of the stria terminalis. The SON, however, was insensitive to these unilateral injections of supraphysiological concentrations of sucrose, a result which is somewhat at variance with the aforementioned electrophysiological results. Although Milton and Paterson (1974) observed ADH release with application of hypertonic NaCl to the SON of cats, it is difficult to draw any conclusions from this study because the molar solution of NaCl was grossly nonphysiological.

Other evidence pertaining to the cerebral location of osmoreceptors is contained in data from investigations in which brain tissue has been ablated. Results from such experiments cannot delineate precise function, but if they are prudently considered, valuable information may be gleaned and particular regions may be definitely excluded from having osmoreceptor function. Forming a background to these experiments were findings in sheep (McKinley et al. 1978) and dogs (Thrasher et al. 1980 b) which suggested that osmoreceptive elements may exist in a brain region lacking the blood-brain barrier. Blood vessels in circumventricular organs such as the subfornical organ, the organum vasculosum of the lamina terminalis (OVLT), the median eminence, and the area postrema lack a blood-brain barrier and therefore these regions have been considered as candidates for housing osmoreceptors. Of these organs, only the OVLT lies within the hypothalamic-preoptic region considered by Jewell and Verney (1957) as the site within which osmoreceptors would be found. Thus, both our group in Melbourne and that of Ramsay and Thrasher in San Francisco have recently studied the feasibility of osmoreceptors being located in the OVLT. Plasma ADH levels were measured in response to systemic hypertonicity before and after ablation of the OVLT in dogs and sheep. The OVLT (or supraoptic crest as it is termed in the older literature) lies in the midline anterior wall of the optic recess of the third cerebral ventricle. When Thrasher et al. (1982 a) ablated the OVLT in the dog they found that although the basal level of ADH in plasma was not altered, the increase in ADH secretion which occurs in response to systemic hypertonicity was greatly attenuated, as was the water-drinking response to this stimulus. Our results in sheep were comparable in that plasma AVP levels in response to intracarotid infusion of hypertonic NaCl (which raised cerebral plasma osmolality by 20 mosm/kg) were substantially reduced by ablation of the OVLT region (McKinley et al. 1982). These studies were extended to examine the effect of such lesions on ADH release to a nonosmotic stimulus such as hemorrhage and the results are presented in Figs. 2 and 3. We found that ablation of the OVLT region greatly reduced ADH secretion in response to a pure hyperosmolar stimulus given over 20 min, but did not reduce ADH secretion to the nonosmotic stimulus of hemorrhage. When the median preoptic nucleus was ablated in addition to the OVLT, there was virtual abolition of ADH secretion in response to hypertonic saline infusion. This nucleus lies immediately dorsal to the OVLT (Fig. 3) and it is possible that a continuum of tissue in the ventral wall of the anterior third ventricle is involved in ADH release to hypertonicity. Although ablation of the OVLT region of sheep caused relatively specific disruption of osmotically mediated ADH release, which is consistent with the notion that osmoreceptors may be located in the OVLT of sheep, this does not prove such a notion, and different methodology will be needed to determine whether this is really the case. Somewhat analogous to these results were earlier finding in goats and rats that ablation of tissue in the anterior wall of the third ventricle depressed the antidiuretic and dipsogenic responses to hypertonicity. The tissue ablated in studies in goats (Andersson et al. 1975; Rundgren et al. 1979) comprised most of the anterior wall of the third ventricle, which along with the OVLT contained median preoptic and medial septal tissue as well as the anterior commissure. In the studies made in rats (Buggy and Johnson 1977; Johnson and Buggy 1978) the anteroventral third ventricle wall was ablated and areas destroyed in addition to the OVLT were the median preoptic nucleus, the periventricular system, and the anterior hypothalamic region. In rats, microinjec-

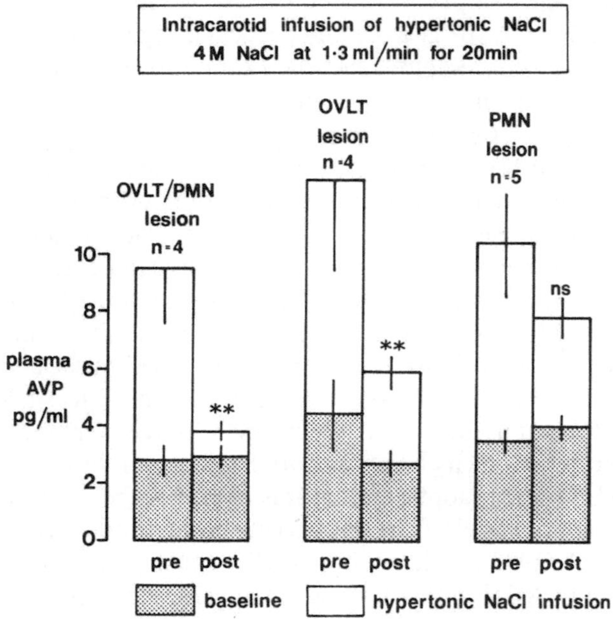

Fig. 2. Effect on plasma [AVP] of infusion of 4 *M* NaCl at 1.3 ml/min for 20 min into one carotid artery with the opposite carotid occluded in sheep with various parts of the anterior wall of the third ventricle ablated. The *hatched bar* gives the plasma [AVP] before the commencement of infusion and the *open bar* after 20 min infusion. Mean ± SEM is shown and ** represents $p < 0.01$ (paired t-test). Responses were measured before (*pre*) and 1–3 months after (*post*) ablating the brain tissue, which included the OVLT or preoptic median nucleus (PMN) separately or both in combination. In four sheep with combined OVLT/PMN lesions and one with OVLT lesion, the effect of hemorrhage of 12.5 ml blood/kg body weight was tested. Plasma [AVP] rose to 121 ± 38 pg/ml in these five animals, a much greater increase than that (6.3 ± 1.3 pg/ml) observed in seven normal animals subjected to this degree of hemorrhage

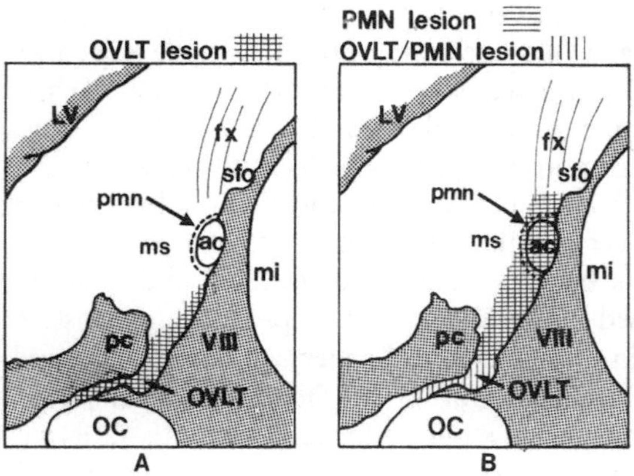

Fig. 3 a, b. The tissue ablation in the three groups of sheep shown in Fig. 2. Diagrams of the midsagittal region of the third ventricle are shown. **a** OVLT lesion (cross-hatching) in which tissue up to 1.5 mm on either side of the coronal midline around the OVLT was ablated but tissue dorsal to this region was left intact. **b** OVLT/PMN lesion (*vertical hatching*) in which the OVLT and PMN were jointly ablated and PMN lesions (*horizontal hatching*) in which the PMN was ablated along with medial septal tissue but the ventral part of the OVLT remained intact. The hatching represents the common tissue damage. *fx*, fornix: *LV*, lateral ventricle; *mi*, massa intermedia; *ms*, medial septum; *oc*, optic chiasma; *pc*, prechiasmatic cisterna; *sfo*, subfornical organ; *VIII*, third ventricle

tion of hypertonic solution into the anteroventral third ventricle resulted in antidiuresis, suggesting that a sensor may be located within this region (Buggy et al. 1979). However, it is difficult to estimate how far such an injection would spread, and the concentrations used were supraphysiological.

It appears to be a consistent finding that ablation of various tissue in the anterior wall of the third ventricle of several species causes large disruption of the osmoregulatory control of ADH release. In all these studies, in goats, rats, dogs, and sheep, the lesions have been confined to midline tissue and the SON and its projection to the neurohypophysis have been spared from damage, yet osmotically stimulated ADH release has been largely and in some instances almost completely disrupted. Accordingly, it seems fair to conclude that the SON cannot to be the *sole* site of osmoreceptors in the brain. While the aforementioned data is consistent with a crucial role of tissue in the OVLT or nerve fibers passing through the OVLT in osmoregulation, further experiments will be necessary to establish whether or not the OVLT is the site of osmoreceptors. It should also be pointed out that ablation of the OVLT in dogs or sheep did not completely abolish increased ADH secretion in response to hypertonicity, which suggests that if the osmoreceptors were in the OVLT, there must also have been other sensors elsewhere involved in osmoregulation.

2.5 Hepatic Sensors Regulating ADH Secretion

The possibility that osmoreceptors in the liver participate in the regulation of ADH secretion was initially put forward by Haberich and co-workers. Although this topic has received far less attention than that of cerebral osmoreceptors, lively debate has taken place on whether such sensors actually exist. The initial evidence supporting the idea of hepatic osmoreceptors has been summarized by Haberich (1968, 1971) and Sawchenko and Friedman (1979). Briefly, Haberich found that increased urine flow occurred in response to infusions of water into the portal vein of the rat, but not the vena cava. Conversely, administration of hypertonic NaCl, mannitol, or glucose into the portal vein selectively reduced urine flow. Other workers have produced evidence consistent with these results (Dennhardt et al. 1971; Adachi et al. 1976). Seemingly contradicting these elegant studies were the findings of Schneider et al. (1970) and Glasby and Ramsay (1974), who were unable to find evidence that infusions of hypotonic solutions or water into the portal vein had any selective effect on urine production in conscious dogs. Unfortunately no direct measurements of plasma ADH levels were made in any of the above studies, and it is also possible that rats and dogs differ in their response to intraportal infusions of hypotonic solutions. Recently, the measurement of plasma AVP concentration has been made in such experiments in anesthetized rats by Baertschi and Vallet (1981), who found not only that superfusion of the portal vein with hypertonic NaCl increased plasma AVP levels, but also that electrical activity in the SON and pituitary stalk fibers increased, indicating greater release of ADH. These workers did not determine whether the responses were due to the increased NaCl concentration per se or to the hyperosmolality. They considered the changes induced in NaCl concentration in portal vein blood in some experiments to be probably within physiological bounds, although the evidence for this is not clear. They also suggested that the sensors were located in the walls of the hepatic portal vein rather than in the liver itself. Although such considerations as anesthesia and the degree of hypertonicity tend to limit the validity of the conclusions drawn, it is obvious that intrahepatic portal vein sensors must be considered as possible physiological regulators of ADH release. Further studies are clearly warranted in order to establish how widespread such sensors may be across different species and also to determine what physiological role they play relative to that of the cerebral osmoreceptors in the regulation of ADH release and water excretion.

In summery, a solid body of evidence is now available in accord with Verney's hypothesis that sensors within the diencephalon detect changes in the effective osmotic pressure of blood and signal the supraopticoneurohypophysial complex to release ADH into the blood stream. Sensors directly detecting altered effective osmotic pressure (osmoreceptors) are probably located somewhere in the anterior hypothalamic median preoptic region, and may be situated at least in part in a brain region lacking the blood-brain barrier, most likely the OVLT. The other favored locus of osmoreceptors is the supraoptic nucleus. In goats and sheep, it seems likely that cerebral sodium sensors additionally play an important role in mediating osmotically induced changes in ADH levels. Osmoreceptors in the liver also appear to play a role in regulating ADH levels, although their importance relative to that of the cerebral sensors remains to be determined.

3 Extracellular Fluid Volume and ADH Secretion

3.1 Theories on the Sensors Involved in Detecting Changes in ECF Volume

As well as the osmoregulatory control of vasopressin secretion, the other major regulation of ADH from a homeostatic point of view comes as a consequence of changes in the blood or ECF volume. If Verney is to be considered as the father of studies of the osmotic regulation of ADH, then Gauer and Henry must surely fill corresponding positions with regard to the study of the volume regulation of ADH. As early as 1951 these investigators proposed the concept of intrathoracic volume sensors regulating renal water excretion and ADH release in response to alterations in the blood volume and filling of the heart (Henry and Gauer 1951; Gauer et al. 1954), and it is considered that these sensors correspond to the type B stetch receptors in the left atrium (Paintal 1973). While there was virtually unanimous acceptance of Verney's view of the osmoreceptors in the two decades following the submission of that theory, the concept of intrathoracic receptors regulating ADH did not enjoy such unquestioned approval and has remained a controversial issue. The other sensors which have been proposed as signalling alterations in ECF volume are the arterial baroreceptors in the carotid sinus and aortic arch, and Share and Levy (1962) postulated that both atrial stretch receptors and baroreceptors were involved in regulating ADH release in response to changed blood volume.

3.2 Atrial Volume Receptors

By using a number of devices, such as negative pressure breathing, immersion, weightlessness, and left atrial distension, to cause engorgement of the intrathoracic circulation and result in diuresis, Gauer, Henry, and fellow workers have proposed that sensors in the heart which respond to altered cardiac filling play a role in regulating ADH secretion. Because cooling of the vagus nerve interfered with diuretic responses, they also suggested that the afferent signal conveying information from the cardiac receptors to the hypothalamic neurosecretory apparatus was carried via the vagus nerve (for review see Gauer and Henry 1976). Despite the fact that numerous investigators (Johnson et al. 1969; Kinney and DiScala 1972) have replicated their original finding that inflation of a balloon in the left atrium, but not in the right atrium (Brennan et al. 1971) increases urinary water output (Henry et al. 1956), arguments have continued over the years as to the mechanism of this diuresis. While it was originally considered that this effect was due to reduction in the circulating levels of ADH, others have challenged this conclusion. These dissenting views have been comprehensively discussed by others (Goetz et al. 1975; Gauer and Henry 1976; Malvin 1979; Schrier et al. 1979), and in summary, the major objections are based on evidence suggesting that alterations in systemic and renal hemodynamics (Lydtin and Hamilton 1964), secretion of a diuretic substance (Ledsome and Mason 1972; Kappagoda et al. 1975), or altered renal nerve activity (Kappagoda et al. 1979) influence urine production when the left atrium is distended. In particular, much debate has centered on whether the diuretic effects of atrial distension are attributable to increases in arterial blood pressure influencing arterial baroreceptor function rather than stimulation of volume receptors in the heart. Additionally, succinct arguments have been advanced which seriously question the conclusion that left atrial receptors are of physiological relevance to the regulation of ADH secretion and bodily water balance. The reader is referred to Goetz et al. (1975) for a comprehensive critical analysis which presents these arguments against the importance of the Henry-Gauer reflex in the regulation of fluid volume. Nevertheless, right up to Dr. Gauer's death in 1979, the original proponents of the concept of atrial receptors continued to supply compelling evidence in favor of this hypothesis at regular intervals (Gauer and Henry 1963; Gauer 1968, Gauer and Henry 1976; Gauer 1980).

The above-mentioned review by Goetz et al. (1975) serves as a useful starting point for our present deliberations, for it was at about this time that a new development began to contribute to considerable clarification of some of these uncertainties. This development was the advent of a number of highly sensitive, specific, and reliable radioimmunoassays for AVP (Skowsky et al. 1974; Robertson et al. 1973; Beardwell 1971). Since the introduction of radioimmunoassay for the measurement of plasma AVP, it is now clear that distension of the left atrium does result in reduced plasma AVP concentration. Inflation of a balloon in the left atrium of dogs was found to cause a reduction in circulating ADH levels and this effect was abolished by extirpation of the pituitary gland (de Torrente et al. 1975). These results were not completely conclusive, however, as the initial vasopressin levels were relatively high (27 pg/ml), probably because the dogs were subjected to anesthesia, laparotomy, and 18 h water deprivation prior to experimentation. Nonetheless, stretching of the left atria of anesthetized dogs and cats has been found to decrease the firing rate of neurons in the SON and PVN, which project to the posterior pituitary (Kannan and Yagi 1978; Koizumi and Yamashita 1978; Menninger 1979). Vagotomy abolished this effect and the results are consistent with the Gauer-Henry hypothesis. More recently, careful studies in conscious dogs have added further evidence verifying that increases in left atrial pressure, rather than intrapulmonary pressure reduce ADH secretion (Schultz et al. 1982). In discussing their results, Schultz et al. stated: "Overall, however, the potency and reproducibility of the left atrial response in the conscious dog is impressive. The multiple hemodynamic, renal and humoral changes that make up this composite left atrial response imply that receptors in the left atrium influence a host of factors that govern cardiovascular function." This conclusion, coming as it does from the "devil's advocate" group of Goetz and associates, is most compelling. The same group were able to abolish the inhibition of ADH secretion and the diuretic response to increased atrial pressure by cardiac denervation. Surprisingly, however, this cardiac denervation did not block the diuresis or reduction of plasma ADH concentration which follows expansion of the circulating blood volume (Fater et al. 1982), and the question still remains of the actual role that the atrial receptors play in the physiological regulation of ADH release. Perhaps the contemporary viewpoint of those sceptical of the importance of atrial receptors in the physiological regulation of ADH is best summarized by the final sentences of Fater et al. (1982): "Our results make it unlikely that alterations in blood volume by the infusion of an isotonic isooncotic solution produce changes in renal function, primarily through a reflex elicited from left atrial receptors. More work is needed to define the precise physiological role of atrial receptors in the regulation of cardiovascular and renal function."

Most evidence advanced in favor of left atrial receptors regulating ADH release independent of baroreceptor function has been derived from experiments involving overfilling of the thoracic circulation (e.g., immersion, atrial distension), but there is some data suggesting that these stretch receptors also respond to reduced filling and initiate increased ADH release in dogs. Although ligation of the inferior vena cava of dogs reduced left atrial pressure and increased ADH secretion and fluid retention (Anderson et al. 1974; Thrasher et al. 1982b), this procedure may have influenced ADH release via alterations in baroreceptor activity, because arterial blood pressure falls (Thrasher et al. 1982b). An interesting study which may clarify this dichotomy was carried out by Yaron and Bennett (1978), who found that inflation of a balloon in the pulmonary artery of conscious dogs not only reduced left atrial pressure, but also increased arterial blood pressure. These workers found that such a procedure increased plasma ADH levels and reduced urine flow, even though arterial pressure rose and plasma osmolality fell due to a simultaneous intravenous infusion of hypotonic saline. They suggested that in the dog, reduction of left atrial pressure can influence vasopressin secretion despite opposing influences from baroreceptors and osmoreceptors. Together with others (Johnson et al. 1969; Zehr et al. 1969) Yaron and Bennett (1978) considered the degree of change in left atrial pressure which they induced by their various experimental strategies to be modest, and the evidence favored a physiological role for left atrial receptors in ADH release in dogs.

3.3 Volume Receptors in Primates

Up to this point, the discussion of atrial receptors has revolved around results obtained almost exclusively in dogs. Recently, however, considerable attention has been given to this question in man and other primates. In man, the evidence for localization of atrial volume receptors is necessarily of an indirect nature. The approach has been to measure changes in vasopressin levels and renal free water clearance during maneuvers which alter intrathoracic blood volume. Indirect evidence favoring involvement of an intrathroracic volume sensor in ADH release in man has come from experiments where subjects were immersed in water up to the neck, which causes translocation of blood into the thoracic circulation (Epstein et al. 1981). Immersion caused a suppression of AVP in plasma with increased urine output. Changes in arterial blood pressure or in the activity of the renin-angiotensin system were probably not the cause of the reduced AVP levels, and the data are consistent with the concept of an intrathoracic receptor regulating AVP release when blood volume is expanded. However, in the situation of underfilling of the circulation, there is little evidence in man to support the concept that sensors other than those responding to alterations in arterial pressure (baroreceptors) mediate ADH release. Nonhypotensive hemorrhage of approximately 10% of blood volume (Goetz et al. 1974; Morton et al. 1975), diuretic-induced fluid loss (Bayliss and De Beer 1981), and hemofiltration in uremic patients (Caillens et al. 1980), all procedures which cause considerable losses of ECF, do not appear to stimulate ADH release until there is sufficient reduction in blood volume to bring about a reduction in arterial blood pressure. Although reduction in intrathoracic blood volume due to tilting to an upright position elevated plasma AVP levels (Davies et al. 1976; Bayliss and Heath 1977), it is possible that decreased blood pressure, nausea, or stress contributed to the elevation, and others have not observed increased ADH release with the assumption of an upright posture (Morton et al. 1975; Kimura et al. 1976). When subjects where treated with a diuretic and then assumed an upright position, a 20% reduction in plasma volume ensued, which resulted in increased ADH levels probably as a consequence of the hypotension which occurred (Kimura et al. 1976). Bayliss et al. (1978) found that application of negative pressure to the lower body, which reduces central blood volume, also stimulated ADH release, but these findings are complicated by episodes of syncope in their subjects. Indeed, in a recent study it was found that application of lower body negative pressure, although causing a significant reduction of 6 mm Hg in central venous pressure, did not induce ADH release till there was a reduction in arterial pressure (Goldsmith et al. 1982), and baroreceptor stimulation has been shown to cause ADH secretion in anesthetized human subjects (Hammer and Engell 1982).

More direct investigations of atrial volume receptors have been made in monkeys. In a series of studies by Gilmore and co-workers, neither a 15% intravascular volume loading (Gilmore et al. 1980) nor left atrial distension (Gilmore and Zucker 1978 a) nor head-out immersion (Gilmore and Zucker 1978 b) were effective in reducing ADH levels. These workers have postulated that atrial stretch receptors are probably less significant in primates, being an adaptation to the frequent changes in posture from upright to recumbent which occur in primates and which would greatly influence left atrial filling even though no overall changes in body fluid volume had occurred. A study using ketamine-anesthetized monkeys found that a 25% increase in plasma volume with isoncotic isotonic fluid reduced ADH levels without increasing blood pressure (Billman et al. 1983), and it was suggested that pentobarbital anesthesia may have influenced the results obtained by Gilmore's group. However, this is not necessarily the case, since Cornish and Gilmore (1982) have shown that unanesthetized monkeys do not respond with diuresis to increased left atrial pressure. With regard to ADH release in response to reduced blood volume in monkeys, Arnauld et al. (1977) subjected rhesus monkeys to hemorrhage of 10%–20% of total blood volume and found that although plasma AVP concentration increased, the increase was related not to the volume of blood withdrawn, but rather to the rate of its withdrawal and to the arterial blood pressure. These workers concluded that ADH release in the unanesthetized monkey was governed by baroreceptors rather than atrial receptors in situations where blood volume fell.

In summary, although atrial sensors cannot be entirely discounted from influencing ADH release during expansion of body fluids, the bulk of evidence obtained in primates argues against a role of atrial receptors in the regulation of ADH release when blood or ECF volume falls. More likely, baroreceptors initiate ADH secretion in primates in response to hypovolemia, and a more detailed discussion of this mechanism now follows.

3.4 Baroreceptors and ADH Secretion

While debate continues on the role of atrial receptors in the regulation of fluid volume, more certainty surrounds the participation of arterial baroreceptors in the regulation of ADH release. Most research concerned with baroreceptor regulation of ADH release has concentrated on mechanisms associated with reduced blood pressure. Initially, the work of Share and Levy (1962) established the important role of baroreceptors in ADH secretion. In a series of studies in anesthetized vagotomized dogs they showed that (a) occlusion of the common carotid arteries increased plasma ADH titers (Share and Levy 1962), (b) carotid sinus pulse pressure was a factor regulating ADH release (Share and Levy 1966), (c) perfusion of the carotid sinus baroreceptors with constant mean and pulse pressure greatly attenuated ADH release in response to hemorrhage (Share 1967), and (d) sectioning of the nerves from the carotid sinus abolished the ADH response to hemorrhage (Share 1968). Data was also obtained in the anesthetized dog which showed that inflation of a balloon in the left atrium could abolish ADH release caused by occlusion of the carotid arteries (Share 1965), and neither vagotomy nor section of the carotid nerves alone inhibited ADH release in response to hemorrhage but did so when carried out jointly. Share and Levy (1962) postulated that both atrial stretch receptors and baroreceptors in the carotid sinus and aortic arch "act in a manner analogous to the function of the baroreceptors in the regulation of arterial blood pressure and exert a tonic inhibition on neurohypophysial ADH release." They also suggested that reduction of this inhibition from vagal and carotid nerve afferents was the mechanism whereby ADH was released in response to hypovolemia. Subsequent studies by other groups using anesthetized

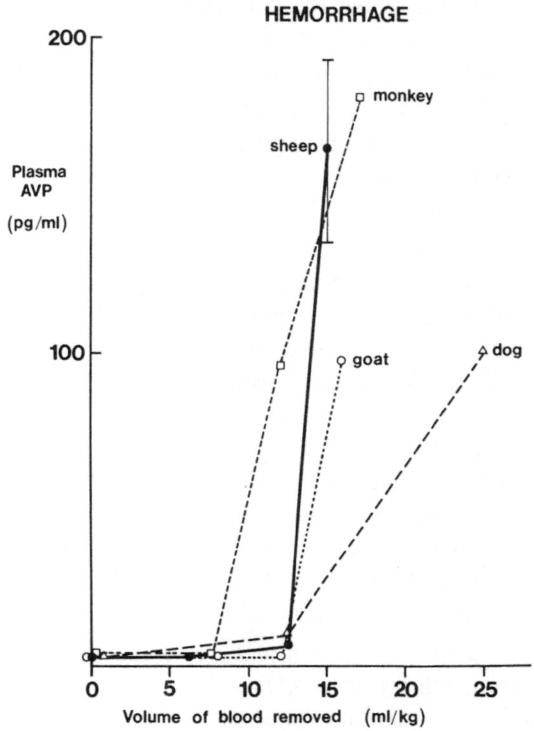

Fig. 4. The relationship between plasma [AVP] and volume of blood withdrawn from the jugular vein of seven unanesthetized sheep (*continuous line*). Blood was progressively withdrawn over 30–40 min until 12.5–15 ml/kg body weight had been removed. Plasma [AVP] was measured by radioimmunoassay and mean ± SEM is shown. For purposes of comparison of species, the results of other workers (*interrupted lines*) have been redrawn from their papers and show the effects of hemorrhage in responding goats (Larsson et al. 1978), the dog (Weitzman et al. 1980), and at a slower rate in monkeys (Arnauld et al. 1977)

dogs (Saito et al. 1969; Thames and Schmid 1979) have produced further evidence in support of this concept. Electrophysiological recordings from neurosecretory SON neurons identified by antidromic stimulation of the neurohypophysis also confirm this idea; occlusion of the carotid artery in rats and dogs has been shown to increase the firing rate of these cells (Kannan and Yagi 1978; Menninger 1979; Koizuma and Yamashita 1978).

The investigations cited above all involved anesthetized animals, and the most common approach to this question in concious animals has been to study the effects of hemorrhage on ADH release. Without exception, graded hemorrhage has resulted in a strikingly similar release of ADH in the species which have been studied, namely monkeys (Arnauld et al. 1977), goats (Larsson et al. 1978), sheep (Fig. 4), and dog (Weitzman et al. 1978a), and by inference in man (Morton et al. 1975). Removal of blood does not increase plasma AVP levels until 10%–15% of the total blood volume has been removed, at which stage a sudden exponential increase in the plasma AVP concentration occurs with increasing blood loss (Fig. 4). In all these studies, it was observed that arterial blood pressure had fallen by the moment when ADH was released. Additionally, it seemed unlikely that this release was related to central venous pressure or reduction of atrial volume (Larsson et al. 1978; Arnould et al. 1977). Since reduction in arterial pressure induced by nitroprusside also induces a large exponentially increasing release of ADH in conscious monkeys (Fumoux et al. 1978), it seems likely that ADH release in response to hemorrhage in most mammals is the result of altered baroreceptor activity. Recent studies in conscious dogs (Schwartz and Reid 1981) and anesthetized rats (Laycock et al. 1979) indicate the ADH release is involved in maintaining arterial blood pressure during mild nonhypotensive hemorrhage. Thus dogs which appear to be most sensitive to changes in atrial pressure for ADH release may also have extremely sensitive baroreceptors which should not be excluded from involvement with the secretion of ADH in response to mild hemorrhage.

With regard to increases in blood pressure, there are few direct studies on the influence of baroreceptors on ADH release when arterial blood pressure increases, but evidence from experiments in which anesthetized dogs were systemically infused with pressor doses of noradrenaline suggest that the resultant increases in free water clearance were due to arterial baroreceptor activation causing an inhibition of ADH release (Schrier et al. 1973; Berl et al. 1974). The same workers were able to show that denervation of aterial baroreceptors or hypophysectomy abolished noradrenaline-induced water diuresis, and since this catecholamine does not readily cross the blood-brain barrier (Oldendorf 1971), it is unlikely it directly affects the activity of cells in the SON. Increased blood pressure also inhibits phasically firing SON neurons (Harris 1979).

In summary, although these data are suggestive of baroreceptor involvement in the inhibition of ADH release in response to increase of arterial blood pressure, direct measurements of decreased ADH levels in conscious animals in response to pressor agents will need to be made before this point is established. With regard to release of ADH in response to hemorrhage, there is good correlation between the fall in arterial blood pressure and increased ADH release. Although this is not unchallengeable proof, it strongly suggests that baroreceptors exert a major influence on neurosecretory cells of the SON during hemorrhage. Along with the study of atrial receptors, consideration of baroreceptor influences have been the major preoccupations of investigators in this field. As an afterthought, one wonders whether the possibility of involvement of extravascular sensors in detection of alterations in ECF has been neglected.

3.5 Interactions Between Volemic and Osmotic Stimuli for ADH Secretion

It is not an uncommon physiological situation for ADH secretion to be simultaneously influenced by alterations in both the ICF volume (osmotic factors) and the ECF volume. Common examples of this are simple dehydration, where hypertonicity and hypovolemia exert combined stimulatory influences on ADH release, and hypertonic NaCl loading, where hypertonicity and hypervolemia exert opposing influences on the neurosecretion of ADH. At one time it was popular to claim that either volemic or osmotic influences were the dominant stimuli for ADH release, with data being produced to show that either osmotic or volemic influences could override other stimuli in regulating ADH secretion. Johnson et al. (1970) recognized that neither system exerted a dominant influence on the release of ADH in sheep, and more recently, with the availability of radioimmunoassays for

AVP, this view has been borne out by the results of detailed investigations (Dunn et al. 1973; Robertson and Athar 1976; Quillen et al. 1983). These investigators found that hypovolemia, although not of sufficient magnitude to induce appreciable ADH release, influenced the ADH release in response to hypertonicity in rats, man, and dogs. Hypovolemia reduced the threshold and increased the gradient of the regression line relating plasma AVP concentration to plasma osmolality. The greater the degree of hypovolemia, the steeper the gradient of this line. Conversely, hypervolemia reduced the steepness of this slope and increased the threshold for osmotic release of ADH. Thus altered volume did not prevent the osmoregulatory functioning of the neurohypophysial system, but adjusted the degree of response to stimulation of the osmoreceptor. In the study in conscious dogs (Quillen and Cowley 1983), left atrial pressure and arterial blood pressure were also recorded, and the modulation by volemic factors of the plasma AVP response to hypertonicity correlated with the left atrial pressure. Quillen and Cowley reasoned that the interactions between osmotic and volemic factors could modify plasma AVP levels by a few picograms per milliliter with consequent alterations in urine loss over several hours for the maintenance of body fluid volume. If, however, hypovolemia is of sufficient magnitude to lower blood pressure, a large increase in plasma AVP concentration to pressor levels may occur, and the relationships mentioned above may not hold.

In a practical sense, the synergistic interaction between osmotic and volemic factors is seen with dehydration, where the plasma AVP levels are greater for a given plasma osmolality than when the osmolality is raised by addition of solute (Robertson and Athar 1976). It has been pointed out by Dunn et al. (1973) that such a mechanism ensures that AVP secretion is maintained during prolonged dehydration, a situation where the neurohypophysial stores of ADH become depleted. They propose that the enhancement of osmotically stimulated ADH release by hypovolemic factors acts as a "booster" mechanism to maintain adequate ADH secretion during extended dehydration. We have observed that dehydration induces a natriuretic response in several species (McKinley et al. 1983 a) which almost certainly lessens the increase in plasma osmolality which would otherwise occur with dehydration (McKinley et al. 1983 b). Thus a booster effect of hypovolemia to enhance the osmotic release of ADH would also help to overcome the drawback of a reduced hyperosmotic stimulus to ADH secretion during dehydration. With regard to the opposite situation of water loading, based on the mechanisms predicted by Quillen and Cowley (1983), hypervolemia would presumably enhance the inhibitory effect of systemic hypotonicity on ADH secretion; however, little investigation has been made into this point because the limit of sensitivity of most assays for AVP is about 1–2 pg/ml and water loading probably leads to lower plasma AVP levels than this. The effects of sodium depletion and sodium loading were studied in man in relation to physiological circumstances where opposing osmotic and volemic factors interact to influence ADH release (Weitzman et al. 1978 b). Although the hypovolemia of sodium depletion appeared to prevent a reduction in plasma AVP concentration in response to the hypotonicity, the plasma osmolality and sodium concentration were only slightly changed by these procedures, indicating that osmolar homeostasis still prevailed.

4 Behavioral Influences on ADH Secretion

When a dehydrated animal is allowed access to water, it drinks and replenishes its volume deficit. There is no longer the dire need for renal water conservation, and plasma AVP levels return to those seen prior to dehydration (Blair-West et al. 1980; Thrasher et al. 1981). While it might reasonably be expected that the increase in blood volume and decrease in blood osmolality due to the subsequent absorption of water from the gastrointestinal tract would be the signals leading to inhibition of ADH release, studies by Thrasher et al. (1981) suggest this is not initially the case. These workers showed that when dehydrated dogs drank water, plasma AVP concentrations fell to reach water-replete levels 15 min after drinking. Inhibition of ADH release occurred very rapidly, before any fall in plasma osmolality was detected. It also occurred when an isotonic saline solution was drunk and when absorption of imbibed water was prevented by allowing it to escape from a gastric fistula. If water was administered into the stomach via the gastric fistula rather than by drinking, the decline in plasma AVP levels was much slower. Inhibition of ADH release appeared to be related to the act of drinking, and Thrasher et al. (1981) proposed that oropharyngeal factors and factors relating to satiety influenced ADH secretion in this situation. We have observed that sheep also reduce plasma ADH levels rapidly when they drink water or slightly hypertonic saline after periods of dehydration, and there is temporarily almost complete cessation of ADH secretion with satiety of drinking before any change in plasma osmolality or plasma volume as measured by plasma protein concentration (McKinley and Congiu, unpublished observations). Electrophysiological recordings in monkeys are consistent with the above findings; the activity in SON cells declines during episodes of drinking (Vincent et al. 1972b). Perhaps this reflex inhibition of ADH release is analogous to the anticipatory potodiuretic reflex described by Nicolaidis (1978) and plays an anticipatory role in the regulation of body fluid homeostasis.

5 Influence of Ambient Temperature on ADH Secretion

The awareness that a diuresis occurs on exposure to cold has long been anecdotally recognized, and there is evidence that cold diuresis may be the result of decreased ADH levels in humans and rats (Segar and Moore 1968; Morgan et al. 1983). On teleological grounds, it would seem advantageous for the conservation of body fluid volume that an increase in ambient and body temperatures should lead to ADH release and reduced urine output, because recognition of increased temperature would be an "early warning" of the evaporative water losses which result from compensatory cooling mechanisms such as sweating and panting. Exposure to high temperatures does increase blood ADH levels in man and the pig (Segar und Moore 1968; Forsling et al. 1976), and ADH may also have a role in thermoregulation during fever (Kasting et al. 1982). The mechanisms underlying changes in ADH release with altered temperature have not been extensively studied, but there is evidence that local heating of preoptic tissue involved in temperature homeostasis will cause ADH release and antidiuresis (Szcepanska-Sadowska 1974), while conversely, preoptic cooling results in a water diuresis probably due to inhibition of ADH secretion (Hayward and Baker 1968). Whether this latter effect is the consequence of altered activity of physiological temperature sensors or results from nonspecific depression by cold of neurons involved in ADH release remains to be determined. Recently, Morgan et al. (1983) obtained evidence that reduced ADH levels and cold diuresis occurred before any reduction in core temperature of rats. They proposed that inhibition of ADH release was secondary to increased blood pressure influencing baroreceptor pathways involved in the neurosecretion of ADH. This does not necessarily infer that in the opposite sense, ADH secretion in response to warmth is also a result of reduced arterial pressure, and there is evidence showing that preoptic cooling can reduce the ADH release induced by high ambient temperature (Forsling et al. 1976). Obviously further studies will be required to clarify the mechanisms involved in this thermal regulation of vasopressin secretion.

6 Afferent Pathways for Neurosecretion of ADH

6.1 Neurotransmitters

If the neurosecretory cells of the SON are not themselves sensors of the ICF or ECF volumes, it is most likely that neural inputs into the SON provide the information. The SON and PVN receive afferent inputs from several brain regions including of the mid- and hindbrain (Loewy 1981) and the anterior wall of the third ventricle (Miselis 1981), regions which may receive signals associated with alterations in the volemic and osmolar state of the animal. This raises the questions: What are the neurotransmitter mechanisms conveying such signals to the SON? Is there neurochemical specificity for the information which comes from the various sensors involved in body fluid homeostasis? There is good histochemical evidence that both cholinergic and adrenergic synapses occur in the SON and PVN (Abrahams et al. 1957; Sladek et al. 1980), and immunohistochemical evidence suggests but does not prove that a number of neuropeptides may also be present in the SON (for review see Hökfelt et al. 1978). With regard to the cholinergic synpases, microinjection of acetylcholine or other cholinomimetics directly into the SON elicits an antidiuresis and ADH release which can be prevented by prior administration of ganglion-blocking drugs (Pickford 1947; Milton and Paterson 1974). Application of noradrenaline at the same sites will also cause ADH secretion (Milton and Paterson 1974). Additionally, several different transmitters and peptides appear to stimulate ADH secretion when administered into the lateral or third cerebral ventricles of a number of species. These agents include acetylcholine and cholinomimetics (Olsson 1970; Bhargava et al. 1972; Kuhn 1974; Hatzikostas et al. 1980), alpha-adrenergic agents (Vandeputte-Van Messom and Peeters 1975; Hisada et al. 1977), dopamine (Bridges et al. 1976; Vandeputte-Van Messom and Peeters 1979), histamine (Bhargava et al. 1973;

Tuomisto et al. 1980; Hoffman and Schmid 1978 a), angiotensin II (Mouw et al. 1971; Keil et al. 1975; Fyhrquist et al. 1979), and bradykinin (Hoffman and Schmid 1978 b). The same substances also cause in vitro release of ADH by isolated hypothalamoneurohypophyseal preparations (Eggena and Thorn 1970; Bridges et al. 1976; Hisada et al. 1977; Sladek et al. 1980). In contrast to these results, intraventricular administration of isoproterenol (Bhargava et al. 1972; Vandeputte-Van Messom and Peeters 1976), noradrenaline (Beal and Bligh 1980; Kimura et al. 1981) opiate peptides (van Wimersma Greidanus et al. 1979; Grossman et al. 1980; Lightman and Forsling 1980), glycine, and GABA (Feldberg and Rocha e Silva 1981) have been found to cause inhibition of ADH release. With the possibility of such a wide spectrum of neurochemical participation in ADH secretion, how do these agents fit into a scheme of the homeostatic regulation of ADH? The methods used to answer this question have usually involved the administration of pharmacological antagonists to the various transmitter agents, and at the moment the results are equivocal. Initially, Bridges and Thorn (1970) suggested that information from the osmoreceptors was conveyed to the SON of rats via a pathway involving at least two synaptic events, one involving a cholinergic receptor and another involving monoamine transmission. In these experiments the antagonist drugs were injected into the bloodstream, and although reserpine, phenoxybenzamine, pempidine, pentolinium, and atropine were found to inhibit osmotically stimulated ADH release, some reservations must exist as to whether these peripherally administered drugs actually act within the CNS, as their passage into the brain may have been inhibited by the blood-brain barrier. Milton and Paterson (1974) applied antagonist doses of the ganglion blockers hexamethonium and pempidine and the adrenergic antagonists phentolamine and propranolol directly onto the SON and were not able to block ADH release to hypertonic saline directly applied to the SON. Unfortunately, the conclusions which might be drawn from these results were compromised, as the osmotic stimulus employed (1 *M* NaCl at the SON) was grossly unphysiological, possibly causing nonspecific depolarization, and the ADH release may not have been initiated by activation of the physiological sensors. The failure of i.c.v. injection of atropine (at doses which blocked the antidiuretic action of i.c.v. carbachol) to inhibit the antidiuretic response to infused hypertonic saline in sheep also suggested that central muscarinic receptors were not involved in osmotically stimulated ADH secretion (Hatzikostas et al. 1980). The difficulty associated with the interpretation of these results is that the site of muscarinic receptors stimulated by i.c.v. carbachol is unknown. While it has often been inferred that ADH release induced by i.c.v. application of various neurotransmitters is due to their action at synaptic junctions within the PVN or SON, there is no direct proof of this. Another factor which adds to the difficulty in making firm conclusions is the possibility that some synaptic sites in the SON are at sufficient distance from the ventricular ependyma to be out of the range of influence of drugs administered into the cerebral ventricles.

Perhaps the most compelling evidence for involvement of cholinergic neurotransmission in the afferent pathway to the SON from osmoreceptors is represented by results obtained from in vitro experiments with supraopticoneurohypophysial explants (Sladek and Joynt 1979) in which nicotinic receptor antagonists (but not adrenergic antagonists) blocked osmotically stimulated ADH release into the bathing medium. These results suggest that acetylcholine may be the neurotransmitter at the SON signalling information from the osmoreceptors, provided these sensors are included in the explanted tissue. In applying these findings to the in vivo release of ADH in response to hypertonicity, it will be necessary to relate the magnitude of ADH release into the medium to physiological levels of secretion in the intact animal. There were also earlier results which showed that the angiotensin antagonist saralasin reduced ADH release by these explants in hypertonic medium (Sladek and Joynt 1980), but these may be misleading. It has subsequently been shown that angiotensin II (AII) generation from renin substrate in the bathing medium, rather than synaptic release of AII following the osmotic stimulus, may have been the reason for this result (Sladek et al. 1982).

In regard to the neurotransmitters coding for alterations in response to changed intravascular volume, there was some evidence that GABA or glycine may exert inhibitory influences on volume-related stimuli for ADH release (Feldberg and Rocha e Silva 1978 a); however, GABA and glycine appear to influence neurons in the ventromedullary region rather than directly influencing neurosecretory neurons of the SON (Feldberg and Rocha e Silva 1978 b). In a recent publication, Blessing et al. (1982) found that ablation by electrolysis or kainic acid of the AI noradrenergic neurons in the caudal ventrolateral medulla resulted in large increases in plasma AVP levels and hypertension in rabbits. They suggested that noradrenergic neurons projecting from the medulla to the SON exert tonic inhibitory influences on ADH secretion, and the results of Kimura et al. (1981) are consistent with this view. In view of the involvement of medullary catecholaminergic neurons in cardiovascular control, it is not difficult to envisage a mechanism whereby increase in AVP secretion in response to reductions in arterial blood pressure is mediated by reduction in the firing rate of the AI noradrenergic projections to the SON. Such a hypothesis awaits experimental testing. Various prostaglandins have been shown to influence renal free water excretion and ADH secretion when administered into the cerebral ventricles (Leksell 1978; Hoffman and Schmid 1979; Ishikawa et al. 1981), and their actions have been linked with central adrenergic mechanisms, but once again the physiological significance of these findings is obscure at present.

6.2 The Renin-Angiotensin System and ADH Secretion

The renal renin-angiotensin system participates in many aspects of cardiovascular and hydromineral balance. In addition to its crucial role in the maintenance of blood pressure and aldosterone secretion during sodium deficiency, this system may also be involved in the regulation of water intake and excretion (Fitzsimons 1980), although the exact extent of this involvement awaits more precise definition. Because hypovolemia is associated with increased secretion of renin by the kidneys and subsequent generation of AI and AII, circulating AII could provide a putative afferent signal to the neurohypophysis for ADH release. However, the physiological significance of such a scheme remains far from certain. Following the initial observations that systemic or i.c.v. infusion of AII could increase plasma ADH levels as measured by bioassay and reduce renal free water clearance in dogs (Bonjour and Malvin 1970; Mouw et al. 1971), other investigators claimed that physiological levels of systemically infused AII were ineffective in eliciting ADH release in dogs (Claybough et al. 1972). More recent results using radioimmunoassay for measurement of plasma AVP do not resolve this dilemma. Although intravenous infusions of AII to produce physiological levels of this octapeptide caused increased AVP secretion in conscious dogs (Ramsay et al. 1978), later studies utilizing the same ADH assay found that the levels of AII in blood needed for ADH release in dogs were probably supraphysiological (Reid et al. 1982). Similarly, Hammer et al. (1980) found that systemic infusion of AII in normal or uremic humans which produced moderate increases of aldosterone levels and arterial blood pressure did not raise plasma AVP concentration. The values given for plasma sodium concentration of the subjects tested seemed on the low side of normal, and it is possible that hypo-osmolality could have reduced their responsiveness to AII. As most of the above-mentioned investigators have pointed out, some reservations must be held with regard to negative findings because it is possible that the hypertensive effect of AII exerted an inhibitory influence on ADH release via baroreceptors. This could override any direct stimulatory effect of AII on ADH secretion, and in physiological situations of hypovolemia with raised blood AII levels no such hypertensive effect occurs. With regard to these physiological situations, some workers have observed correlations between the levels of AII or renin and AVP in plasma (Yamaguchi et al. 1980; Weitzman et al. 1978 a), while others do not support such an association (Kimura et al. 1976; Morton et al. 1977; Share 1979). Similarly, blockade of AII receptors by saralasin or prevention of AII formation by inhibition of the action of converting enzyme has produced varied results. Although a converting-enzyme inhibitor (CEI) blocked ADH release in response to hemofiltration in uremic patients (Caillens et al. 1980) and intraventricular saralasin suppressed plasma AVP levels in dehydrated rats (Yamaguchi et al. 1980) and sodium-depleted dogs (Brooks and Malvin 1979), it did not block ADH release in response to hemorrhage (Morton et al. 1977). Thus it is at the moment difficult to find a consensus on the physiological role of the renal renin-angiotensin system in the regulation of ADH release.

This subject is still further complicated by the possibility that a cerebral renin-angiotensin system (Ganten et al. 1971; Ganten et al. 1976; Phillips 1978) participates in the regulation of ADH secretion. Consistent with this view are findings that AII does not pass the blood-brain or blood-liquor barriers (Ganten et al. 1971; Ganten et al. 1976) but has been located by immunocytochemical methods in the vicinity of the SON (Kilcoyne et al. 1980). Even though AII does not cross the blood-brain barrier, it stimulates neurosecretory cells of the SON when directly applied to these neurons by iontophoresis (Nicoll and Barker 1971), and i.c.v. administration of renin or AII elicits potent release of ADH and antidiuresis (Mouw et al. 1971; Keil et al. 1975; Malayan and Reid 1976; Yamaguchi et al. 1980). In addition, i.c.v. administration of CEI or saralasin blocks ADH release or antidiuresis in some physiological situations (Brooks and Malvin 1979; Yamaguchi et al. 1980), and AVP alters converting enzyme activity in the SON and PVN (Saavedra and Chevillard 1982).

While such data suggest that a cerebral renin-angiotensin system exists and participates in ADH regulation, this concept is still controversial and some aspects are clouded in uncertainty: (a) the actual stimuli which "switch on" the putative cerebral renin-angiotensin system, and (b) whether AII is a neurotransmitter or whether it is present in sufficient amounts in tissue to modulate neural activity. The conclusion of Share (1979) that more research is needed regarding the role of this system in the regulation of ADH release is still valid.

7 ADH Secretion in the Fetus

Studies into the ontogeny of ADH release in response to intracellular or extracellular volume depletion have been made during the past 5 years, with the experimental subjects having almost exclusively been sheep. Leake et al. (1979) have shown that AVP does not cross the placenta, and thus the plasma AVP levels measured in the fetal sheep depend on the balance between rate of secretion and rate of removal of AVP by the fetus.

Because plasma AVP concentration increases in the ovine fetus in response to infusion of hypertonic NaCl (Weitzman et al. 1978c; Leake et al. 1979), hemorrhage (Robillard et al. 1979), or hypotension due to administration of sodium nitroprusside (Rose et al. 1981) at 100–145 days of gestation (the third trimester of pregnancy in sheep), it is probable that such stimuli induce release of ADH in the fetus at this stage of development. Obstruction of the vena cava also caused increased plasma AVP levels in 118- to 134-day-old fetal sheep, provided blood pressure fell (Wood et al. 1982). There is a tendency for a progressive increase in responsiveness to a hypotensive stimulus from 100 days of gestation onwards (Rose et al. 1981), whereas there is evidence that the osmoreceptor system is fully functional at 100 days of gestation (Weitzman et al. 1978c). In view of recent findings showing that maternal dehydration caused lower plasma AVP levels for a given plasma osmolality in 107–119-day-old fetuses than in those at 126–144 days of development (Bell et al. 1983), some doubt must exist as to the degree of maturation of the osmoreceptor system early in the third trimester. However, it is possible that this result reflects slower development of the system which responds to ECF volume depletion. It also appears that the fetal kidney, although responding to increased circulating ADH levels by reducing renal free water clearance throughout the final third of gestation, becomes more responsive to ADH as parturition approaches (Wintour et al. 1982); however, this responsiveness is still considerably less than that of the adult sheep (Robillard and Weitzman 1980). The patterns of development of ADH release and its action on the kidney in the fetus contribute to a complex situation in which its internal evironment (blood and ECF volume and compositon) and external environment (amniotic fluid volume and compositon) are homeostatically regulated during gestation.

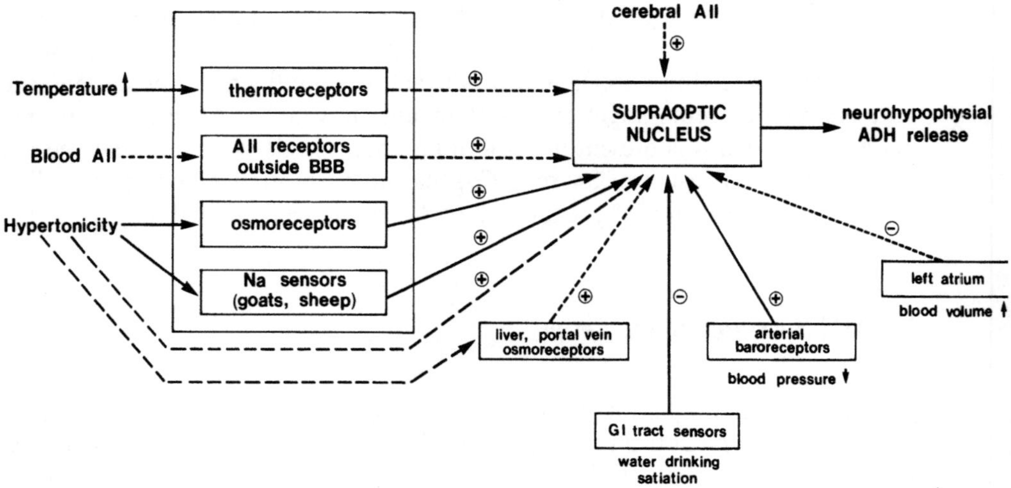

Fig. 5. A summary of factors which may influence ADH release in mammals. *Solid lines* indicate those factors and pathways likely to be physiologically significant in regulating ADH secretion, *interrupted lines* denote pathways of a more speculative nature which need further confirmation. *BBB*, blood brain barrier; *GI*, gastrointestinal

8 Concluding Remarks

The past decade has seen considerable growth in our knowledge of the patterns of secretion of ADH which occur in response to both osmolar and volemic stimuli, and the factors involved are summarized in Fig. 5. This growth has without doubt resulted partly from the widespread use of radioimmunoassays to measure AVP concentration in blood, and the knowledge has also contributed to advances in the clinical treatment of disturbances of body fluid and electrolyte metabolism. This subject is beyond the scope of this article, and the reader is referred to excellent clinical reviews by Robertson (1977), Schrier und Goldberg (1980), Schrier and Bichet (1981), and Robertson et al. (1982). With regard to the future, recent technological advances in the fields of peptide chemistry, molecular biology, and neural and functional mapping of the central nervous system, and their application to this field, bode well for continued expansion of our knowledge of the neuroanatomic and neurochemical pathways which subserve the release of ADH in order to preserve body fluid volume and composition.

References

Abrahams VC, Koelle GB, Smart P (1957) Histochemical demonstration of cholinesterase in the hypothalamus of the dog. J Physiol (Lond) 139:137–144

Adachi A, Niijima A, Jacobs HL (1976) An hepatic osmoreceptor mechanism in the rat: electrophysiological and behavioral studies. Am J Physiol 231:1043–1049

Anderson RG, Cadnapaphornchai P, Harbottle JA, McDonald KM, Schrier KW (1974) Mechanism of effect of thoracic inferior vena cava constriction on renal water excretion. J Clin Invest 54:1473–1479

Andersson B (1977) Regulation of body fluids. Ann Rev Physiol 39:185–200
Andersson B (1978) Regulation of water intake. Physiol Rev 58:582–603
Andersson B, Olsson K, Warner RG (1967) Dissimilarities between the central control of thirst and the release of antidiuretic hormone (ADH). Acta Physiol Scand 64:407–417
Andersson B, Leksell L, Lishajko F (1975) Perturbations in fluid balance induced by medially placed forebrain lesions. Brain Res 99:262–275
Andersson B, Leksell LG, Lishajko F, Rundgren M (1975) Cerebral sodium-noradrenaline interaction: dipsogenic, antidiuretic and natriuretic effects. Acta Physiol Scand 102:254–256
Arieff AI, Guisado R (1976) Effects on the central nervous system of hypernatremic and hyponatremic states. Kidney Int 10:104–116
Arnauld EP, Czernichow F, Fumoux F, Vincent JD (1977) The effects of hypotension and hypovolaemia on the liberation of vasopressin during haemorrhage in the unanesthetized monkey (*Macaca mulatta*). Pfluegers Arch 371:193–200
Baertschi AJ, Vallet PG (1981) Osmosensitivity of the hepatic portal vein area and vasopressin release in rats. J Physiol (Lond) 315:217–230
Baldwin BA, Bell FR (1963) The anatomy of the central circulation of the sheep and the ox. The dynamic distribution of the blood supplied by the carotid and vertebral arteries to cranial regions. J Anat (Lond) 97:203–215
Bayliss PH, DeBeer FC (1981) Human plasma vasopressin response to potent loop-diuretic drugs. Eur J Clin Pharmacol 20:343–346
Bayliss PH, Heath DA (1977) Influence of presyncope and postural change upon plasma arginine vasopressin concentration in hydrated and dehydrated man. Clin Endocrinol 7:79–83
Bayliss PH, Robertson GL (1980) Plasma vasopressin response to hypertonic saline infusion to assess posterior pituitary function. J R Soc Med 73:255–260
Bayliss PH, Stockley RA, Heath DA (1978) Influence of lower body negative pressure upon arginine vasopressin release. Clin Endocrinol 9:89–95
Beal AM, Bligh J (1980) Diuretic effect of intraventricular and intravenous infusion of noradrenaline in conscious sheep. Q J Exp Physiol 65:321–333
Beardwell CG (1971) Radioimmunoassay of arginine vasopressin in human plasma. J Clin Endocrinol Metab 33:254–260
Bell RJ, Congiu M, Hardy KJ, Wintour EM (1984) Gestation-dependent aspects of the response of the ovine fetus to the osmotic stress induced by maternal water deprivation. Q J Exp Physiol 69:108–112
Berl T, Cadnapaphornchai P, Harbottle JA, Schrier RW (1974) Mechanism of suppression of vasopressin during alpha-adrenergic stimulation with norepinephrine. J Clin Invest 53:219–227
Bhargava KP, Kulshrestha VK, Skrivastava YP (1972) Central cholinergic and adrenergic mechanisms in the release of antidiuretic hormone. Br J Pharmacol 44:617–627
Bhargava KP, Kulshrestha VK, Santhakumari G, Srivastava YP (1973) Mechanism of histamine-induced antidiuretic response. Br J Pharmacol 47:700–706
Bie P (1976) Studies of cerebral osmoreceptors in anesthetized dogs: the effect of intravenous and intracarotid infusion of hyper-osmolar sodium chloride solutions during sustained water diuresis. Acta Physiol Scand 96:306–318
Bie P (1980) Osmoreceptors, vasopressin and control of renal water excretion. Physiol Rev 60:961–1048
Bie P, Peterson TV, Share L, Gilmore JP (1982) Osmotic control of plasma vasopressin in anesthetized dogs. Acta Physiol Scand 114:37–43
Billman GE, Keyl MJ, Dickey DT, Kem DC, Keil LC, Stone HL (1983) Hormonal and renal response to plasma volume expansion in the primate *Macaca mulatta*. Am J Physiol 244:H201–H205
Blaine EH, Denton DA, McKinley MJ, Weller S (1975) A central osmosensitive receptor for renal sodium excretion. J Physiol (Lond) 244:497–509
Blair-West JR, Bobik A, Brook AH, Esler MD, Gibson A, Morris M, McKinley MJ, Pullan PT (1980) Renin ADH and the kidney: a congeries of conundrums. Prog Biochem Pharmacol 17:20–28

Blessing WW, Sved AF, Reis DJ (1982) Destruction of noradrenergic neurons in rabbit brain stem elevates plasma vasopressin, causing hypertension. Science 217:661–663
Bonjour JP, Malvin RL (1970) Stimulation of ADH release by the renin-angiotensin system. Am J Physiol 218:1555–1559
Brennan LA Jr, Malvin RL, Jochim KE, Roberts DE (1971) Influence of right and left atrial receptors on plasma concentrations of ADH and renin. Am J Physiol 221:273–278
Bridges TE, Thorn NA (1970) The effect of autonomic blocking agents on vasopressin release in vivo induced by osmoreceptor stimulation. J Endocrinol 48:265–276
Bridges TE, Hillhouse EM, Jones MT (1976) The effect of dopamine on neurohypophysial hormone release in vivo and from the rat neural lobe and hypothalamus in vitro. J Physiol (Lond) 260:647–666
Brooks VL, Malvin RL (1979) An intracerebral, physiological role for angiotensin: effects of central blockade. Fed Proc 38:2272–2275
Brooks C, Koizumi K, Zeballos GA (1966) A study of factors controlling activity of neurons within the paraventricular, supraoptic and ventromedian nuclei of the hypothalamus. Acta Physiol Lat Am 16:83–96
Brownstein MS, Russell JT, Gainer H (1980) Synthesis, transport and release of posterior pituitary hormones. Science 207:373–378
Buggy J, Johnson AK (1977) Preoptic-hypothalamic periventricular lesions: thirst deficits and hypernatremia. Am J Physiol 233:R44–R52
Buggy J, Hoffman WE, Phillips MI, Fisher AE, Johnson AK (1979) Osmosensitivity of rat third ventricle and interactions with angiotensin. Am J Physiol 236:R75–R82
Caillens H, Pruszcynski W, Meyrier A, Ang K, Rousselet F, Ardaillou R (1980) Relationship between change in volemia at constant osmolality and plasma antidiuretic hormone. Miner Electrolyte Metab 4:161–171
Claybaugh JR, Share L, Shimizu K (1972) The inability of infusions of angiotensin to elevate the plasma vasopressin concentration in the anesthetized dog. Endocrinology 90:1647–1652
Cornish KG, Gilmore JP (1982) Increased left atrial pressure does not alter renal function in the conscious primate. Am J Physiol 243:R119–R124
Davies R, Slater JDH, Forsling ML, Payne N (1976) The response of arginine vasopressin and plasma renin to postural change in normal man with observations on syncope. Clin Sci Mol Med 51:267–274
Dennhardt R, Ohm WW, Haberich FJ (1971) Die Ausschaltung der Leberäste des N. vagus an der wachen Ratte und ihr Einfluß auf die hepatogene Diurese – indirekter Beweis für die afferente Leitung der Leber-Osmoreceptoren über den N. vagus. Pfluegers Arch 328:561–56
De Torrente A, Robertson GL, McDonald KM, Schrier RW (1975) Mechanism of diuretic response to increased left atrial pressure in the anesthetized dog. Kidney Int 8:355–361
Dietz JR, Bie P, Gilmore JP, Share L, Zucker IH (1982) The relation between carotid solute concentration and renal water excretion in conscious dogs. Acta Physiol Scand 114:45–51
Dunn FL, Brennan IJ, Nelson AE, Robertson GL (1973) The role of blood osmolality and volume in regulating vasopressin secretion in the rat. J Clin Invest 52:3212–3219
Eggena P, Thorn NA (1970) Vasopressin release from the rat supraoptico-neurohypophysial system in vitro in response to hypertonicity and acetylcholine. Acta Endocrinol 65:442–452
Epstein M, Preston S, Weitzman RE (1981) Isoosmotic central blood volume expansion suppresses plasma arginine vasopressin in normal man. J Clin Endocrinol Metab 52:256–262
Eriksson L (1974) Effect of lowered CSF sodium concentration on the central control of fluid balance. Acta Physiol Scand 91:61–68
Eriksson L, Fernandez O, Olsson K (1971) Differences in the antidiuretic response to intracarotid infusions of various hypertonic solutions in the conscious goat. Acta Physiol Scand 83:554–562

Farini A (1913) Hypophyseal apotherapy in diabetes insipidus. Br Med J Nov 29:76 [Abstr]
Fater DC, Schultz HD, Sundet WD, Mapes JS, Goetz KL (1982) Effects of left atrial stretch in cardiac-denervated and intact conscious dogs. Am J Physiol 242:H1056–H1064
Feldberg W, Rocha e Silva M (1978a) GABA and glycine inhibit vasopressin release to carotid occlusion. J Physiol (Lond) 289:43P–44P
Feldberg W, Rocha e Silva M (1978b) Vasopressin release produced in anaesthetized cats by antagonists of gamma-aminobutyric acid and glycine. Br J Pharmacol 62:99–106
Feldberg W, Rocha e Silva M (1981) Inhibition of vasopressin release to carotid occlusion by gamma-aminobutyric acid and glycine. Br J Pharmacol 72:17–24
Fitzsimons JT (1980) Angiotensin stimulation of the central nervous system. Rev Physiol Biochem Pharmacol 87:117–167
Forsling ML (1977) Antidiuretic hormone. Annual research reviews vol 2. Churchill Livingstone, Edinburgh
Forsling ML, Ingram DL, Stanier MW (1976) Effects of various ambient temperatures and of heating and cooling the hypothalamus and cervical spinal cord on antidiuretic hormone secretion and urinary osmolality in pigs. J Physiol (Lond) 257:673–686
Frank E (1910) Über Diabetes Insipidus als Zeichen gesteigerter Funktion des Hinterlappens der Hypophysis. Berlin Klin Wochenschr 47:1257
Fumoux F, Czernichow P, Arnauld E, DuPont J, Vincent JD (1978) Effect of hypertension induced by sodium nitrocyanoferrate (111) on the release of arginine-vasopressin in the unanaesthetized monkey. J Endocrinol 78:449–450
Fyhrquist F, Eriksson L, Wallenius M (1979) Plasma vasopressin in conscious goats after cerebroventricular infusions of angiotensin, sodium chloride and fructose. Endocrinology 104:1091–1095
Ganten D, Marquez-Julio A, Granger P, Hayduk K, Karsunsky KP, Boucher R, Genest J (1971) Renin in dog brain. Am J Physiol 221:1733–1737
Ganten D, Hutchinson JS, Schelling P, Ganten U, Fischer M (1976) The iso-renin-angiotensin systems in extrarenal tissue. Clin Exp Pharmacol Physiol 3:103–126
Gauer OH (1968) Osmocontrol versus volume control. Fed Proc 27:1132–1136
Gauer OH (1980) Control neuro-endocrinien du volume plasmatique. J Physiol (Paris) 76:373–379
Gauer OH (1963) Circulatory basis of fluid volume control. Physiol Rev 43:423–481
Gauer OH, Henry JP (1976) Neurohormonal control of plasma volume. In: Guyton AC, Cowley AW (eds) Cardiovascular Physiology II. University Park, Baltimore, pp 145–190 (International review of physiology, vol 9)
Gauer OH, Henry JP, Sieker HO, Wendt WE (1954) The effect of negative pressure breathing on urine flow. J Clin Inv 33:287–296
Gilmore JP, Zucker IH (1978a) Failure of left atrial distension to alter renal function in the nonhuman primate. Circ Res 42:267–270
Gilmore JP, Zucker IH (1978b) Contribution of vagal pathways to the renal responses of head-out immersion in the nonhuman primate. Circ Res 42:263–267
Gilmore JP, Zucker IH, Ellington MJ, Richards MA, Share L (1980) Failure of acute intravascular volume expansion to alter plasma vasopressin in the nonhuman primate *Macaca fascicularis*. Endocrinology 106:979–982
Glasby MA, Ramsay DJ (1974) Hepatic osmoreceptors? J Physiol (Lond) 243:765–776
Goetz KL, Bond GC, Smith WE (1974) Effect of moderate hemorrhage in humans on plasma ADH and renin. Proc Soc Exp Biol Med 145:277–280
Goetz KL, Bond GC, Bloxham DD (1975) Atrial receptors and renal function. Physiol Rev 55:157–205
Goldsmith SR, Francis GS, Cowley AW, Cohn JN (1982) Response of vasopressin and norepinephrine to lower body negative pressure in humans. Am J Physiol 243:H970–H973
Gottschalk CW (1964) Osmotic concentration and dilution of the urine. Am J Med 36:670–685

Grossman A, Besser GM, Milles JJ, Baylis PH (1980) Inhibition of vasopressin release in man by an opiate peptide. Lancet 2:1108–1110
Haberich FJ (1968) Osmoreception in the portal circulation. Fed Proc 27:1137–1141
Haberich FJ (1971) Osmoreceptors in the portal circulation and their significance for the regulation of water balance. Triangle 10:123–130
Haller EW, Wakerley JB (1980) Electrophysiological studies of paraventricular and supraoptic neurons recorded in vitro from slices of rat hypothalamus. J Physiol (Lond) 302:347–362
Hammer M, Engell HC (1982) Episodal secretion of vasopressin in man. Acta Endocrinol 101:517–523
Hammer M, Olgaard K, Madsen S (1980) The inability of AII infusions to raise plasma vasopressin levels in haemodialysis patients. Acta Endocrinol 95:422–426
Harris MC (1979) Responses of supraoptic neurones to chemoreceptor and baroreceptor stimulation in rats. J Physiol (Lond) 287:25P–26P
Hatzikostas S, Denton DA, McKinley MJ, Weisinger RS (1980) Central cholinergic receptors and antidiuresis due to hypertonicity in sheep. Neuroendocrinology 30:275–279
Hayward JN, Baker MA (1968) Diuretic and thermoregulatory responses to preoptic cooling in the monkey. Am J Physiol 214:843–850
Hayward JN, Jennings DP (1973) Osmosensitivity of hypothalamic magnocellular neuroendocrine cells to intracarotid hypertonic D-glucose in the waking monkey. Brain Res 57:467–472
Henry JP, Gauer OH (1951) Certain haemodynamic factors concerned with control of blood volume. Fed Proc 10:62
Henry JP, Gauer OH, Reeves JL (1956) Evidence of the atrial location of receptors influencing urine flow. Circ Res 4:85–90
Hisada S, Fusimoto S, Kamiya T, Endo Y, Tsushima H (1977) Antidiuresis of centrally administered amines and peptides and release of antidiuretic hormone from isolated rat neurohypophysis. Japan J Pharmacol 27:153–161
Hökfelt T, Elde T, Fuxe K, Johansson O, Ljundahl A, Goldstein M, Luft R, Efendic S, Nilsson G, Terenius L, Ganten D, Jeffcoate SL, Rehfeld J, Said S, Perez de la Mora M, Possani L, Tapia R, Teran L, Palacios R (1978) Aminergic and peptidergic pathways in the nervous system with special reference to the hypothalamus. In: Reichlin S, Baldessavini RJ, Martin JB (eds) The hypothalamus. Raven, New York, pp 69–135
Hoffman WE, Schmid PG (1978a) Cardiovascular and antidiuretic effects of central histamine. Life Sci 22:1709–1714
Hoffman WE, Schmid PG (1978b) Separation of pressor and antidiuretic effects of intraventricular bradykinin. Neuropharmacology 17:999–1002
Hoffman WE, Schmid PG (1979) Cardiovascular and antidiuretic effects of central prostaglandin E2. J Physiol (Lond) 288:159–169
Ishikawa S, Saito T, Yoshida S (1980) Effects of glucose and sodium chloride on the release of vasopressin in response to angiotensin II from the guinea pig hypothalamo-neurohypophysial complex in organ culture. Neuroendocrinology 31:365–368
Ishikawa S, Saito T, Yoshida S (1981) The effect of prostaglandins on the release of arginine vasopressin from the guinea pig hypothalamo-neurohypophyseal complex in organ culture. Endocrinology 108:193–198
Jewell PA, Verney EB (1957) An experimental attempt to determine the site of the neurohypophysial osmoreceptors in the dog. Philos Trans R Soc London Ser B 240:197–324
Johnson AK, Buggy J (1978) Periventricular preoptic hypothalamus is vital for thirst and normal water economy. Am J Physiol 234:R122–R129
Johnson JA, Moore WW, Segar WE (1969) Small changes in left atrial pressure and plasma antidiuretic hormone titers in dogs. Am J Physiol 217:210–214
Johnson JA, Zehr JE, Moore WW (1970) Effects of separate and concurrent osmotic and volume stimuli on plasma ADH in sheep. Am J Physiol 218:1273–1280
Kannan H, Yagi K (1978) Supraoptic neurosecretory neurons: evidence for the existence of converging inputs both from carotid baroreceptors and osmoreceptors. Brain Res 145:385–390

Kappagoda CT, Linden RJ, Snow HM, Whitaker EM (1975) Effect of destruction of the posterior pituitary on the diuresis from left atrial receptors. J Physiol (Lond) 244:757–770
Kappagoda CT, Linden RJ, Sreehan N (1979) The role of renal nerves in the diuresis and natriuresis caused by stimulation of atrial receptors. J Physiol 287:17P–18D
Kasting NW, Veale WL, Cooper KE (1982) Vasopressin: a homeostatic effector in the febrile process. Neurosci Biobehav Rev 6:215–222
Keil LC, Summy-Long J, Severs WB (1975) Release of vasopressin by angiotensin II. Endocrinology 96:1063–1065
Kilcoyne MM, Hoffman DL, Zimmerman EA (1980) Immunocytochemical localization of angiotensin II and vasopressin in rat hypothalamus: evidence for production in the same neuron. Clin Sci 59:575–605
Kimura T, Minai K, Matsui K, Mouri T, Sato T, Yoshinaga K, Hoshi T (1976) Effect of various states of hydration on plasma ADH and renin in man. J Clin Endocrinol Metab 42:79–87
Kimura T, Share L, Wang BC, Crofton JT (1981) The role of central adrenoreceptors in the control of vasopressin release and blood pressure. Circ Res 42:263–267
Kinney MJ, DiScala VA (1972) Renal clearance studies of effect of left atrial distension in the dog. Am J Physiol 22:1000–1003
Klisiecki A, Pickford M, Rothschild P, Verney EB (1933) The absorption and excretion of water by the mammal. Proc R Soc Lond Ser B 112:496–521
Koizumi K, Yamashita H (1978) Influence of atrial stretch receptors on hypothalamic neurosecretory neurones. J Physiol (Lond) 285:341–358
Koizuma K, Ishikawa T, Brooks CMcC (1964) Control of activity of neurons in the supraoptic nucleus. J Neurophysiol 27:878–892
Kuhn ER (1974) Cholinergic and adrenergic release mechanisms for vasopressin in the male rat: a study with injections of neurotransmitters and blocking agents into the third ventricle. Neuroendocrinology 16:255–264
Land H, Schütz G, Schmale H, Richter D (1982) Nucleotide sequence of cloned cDNA encoding bovine arginine vasopressin neurophysin II precursor. Nature 295:299–303
Larsson B, Olsson K, Fyhrquist F (1978) Vasopressin release induced by haemorrhage in the goat. Acta Physiol Scand 104:309–317
Laycock JF, Penn W, Shirley DG, Walter SJ (1979) The role of vasopressin in blood pressure regulation immediately following acute haemorrhage in the rat. J Physiol (Lond) 296:267–275
Leake RD, Weitzman RE, Effros RM, Siegal SR, Fisher DA (1979) Maternal fetal osmolar homeostasis: fetal posterior pituitary autonomy. Pediatr Res 13:841–844
Ledsome JR, Mason JM (1972) The effect of vasopressin on the diuretic response to left atrial distension. J Physiol (Lond) 221:427–440
Leksell LG (1978) Effects on fluid balance induced by non-febrile intracerebroventricular infusions of PGE_2, $PGF_{2\alpha}$, and arachidonic acid in the goat. Acta Physiol Scand 104:225–231
Leksell LG, Lishajko F, Rundgren M (1976) Negative water balance induced by intracerebroventricular infusion of deuterium. Acta Physiol Scand 97:142–144
Leksell LG, Congiu M, Denton DA, Fei DTW, McKinley MJ, Tarjan E, Weisinger RS (1981) Influence of mannitol-induced reduction in CSF Na on nervous and endocrine mechanisms involved in the control of fluid balance. Acta Physiol Scand 112:33–40
Leng G (1980) Rat supraoptic neurons: the effects of locally applied hypertonic saline. J Physiol (Lond) 304:405–414
Leng G, Mason WT, Dyer RG (1982) The supraoptic nucleus as an osmoreceptor. Neuroendocrinology 34:75–82
Leusen I, Lacroix E (1961) Changes in osmolarity in the cerebral ventricles and diuresis. Endocrinology 68:719–721
Lightman SL, Forsling ML (1980) The effect of the methionine enkephalin analogue DAMME on the vasopressin response to tilt in man. Clin Sci 59:501–503
Loewy AD (1981) Descending pathways to sympathetic and parasympathetic preganglionic neurons. J Auton Nerv Syst 3:265–276

Lydtin H, Hamilton W (1964) Effect of acute changes in left atrial pressure on urine flow in unanaesthetized dogs. Am J Physiol 207:530–536

Macfarlane WV, Howard B (1972) Comparative water and energy economy of wild and domestic mammals. In: Maloiy GMO (ed) Comparative physiology of desert animals. Academic, London, pp 261–296

Malayan SA, Reid IA (1976) Antidiuresis produced by injection of renin into the third cerebral ventricle of the dog. Endocrinology 92:329–335

Malmo RB, Malmo HP (1979) Responses of lateral preoptic neurons in the rat to hypertonic sucrose and NaCl. Electroencephalogr Clin Neurophysiol 46:401–408

Malvin RL (1979) Carnivores. In: Maloiy GMO (ed) Comparative physiology and osmoregulation in animals. Academic, London, pp 145–184

McKinley MJ, Denton DA, Weisinger RS (1978) Sensors for antidiuresis and thirst – osmoreceptors or CSF sodium detectors? Brain Res 141:89–103

McKinley MJ, Olsson K, Fyhrquist F, Liljekvist E (1980) Transient vasopressin release and thirst in response to prolonged intracerebroventricular infusions of hypertonic mannitol in saline. Acta Physiol Scand 109:427–431

McKinley MJ, Congiu M, Denton DA, Park RG, Smith MH, Weisinger RS, Wright RD (1982) Ablation of the organum vasculosum of the lamina terminalis (OVLT) of sheep disrupts osmoregulatory but not haemorrhage-induced vasopressin secretion. Proc Austral Physiol Pharmacol Soc 13:185P

McKinley MJ, Denton DA, Nelson JF, Weisinger RS (1983a) Dehydration induces sodium depletion in rats, rabbits and sheep. Am J Physiol 245:R287–R292

McKinley MJ, Denton DA, Park RG, Weisinger RS (1983b) Cerebral involvement in dehydration-induced natriuresis. Brain Res 263:340–343

Menninger RP (1979) Effects of carotid occlusion and left atrial stretch on supraoptic neurosecretory cells. Am J Physiol 237:R63–R67

Milton AS, Paterson AT (1974) A microinjection study of the control of antidiuretic hormone release by the supraoptic nucleus of the hypothalamus in the cat. J Physiol (Lond) 241:607–628

Miselis R (1981) The efferent projections of the subfornical organ of the rat: a circumventricular organ within a neural network subserving water balance. Brain Res 230:1–23

Morgan ML, Anderson RJ, Ellis MA, Berl T (1983) Mechanisms of cold diuresis in the rat. Am J Physiol 244:F210–F216

Morton JJ, Padfield PL, Forsling ML (1975) A radioimmunoassay for plasma arginine-vasopressin in man and dog; application to physiological and pathophysiological states. J Endocrinol 65:411–424

Morton JJ, Semple PF, Ledingham IMcA, Stuart B, Tehrani MA, Garcia AR, McGarrity (1977) Effect of angiotensin-converting enzyme inhibitor on the plasma concentrations of angiotensin I, angiotensin II, and arginine vasopressin in the dog during hemorrhagic shock. Circ Res 41:301–308

Mouw D, Bonjour JP, Malvin RL, Vander A (1971) Central action of angiotensin in stimulating ADH release. Am J Physiol 220:239–242

Nicolaidis S (1978) Role des reflexes anticipateurs oro-vegetatifs dans la regulation hydrominerale et energetique. J Physiol (Paris) 74:1–19

Nicoll RA, Barker JL (1971) Excitation of supraoptic neurosecretory cells by Angiotensin II. Nature 233:172–173

Noble R, Wakerley JB (1982) Behaviour of physically active supraoptic neurones in vitro during osmotic challenge with sodium chloride or mannitol. J Physiol (Lond) 327:41P

O'Connor WJ, Verney EB (1942) The effect of removal of the posterior lobe of the pituitary on the inhibition of water diuresis by emotional stress. J Exp Physiol 31:393–408

Oldendorf WH (1971) Brain uptake of radiolabeled amino acids, amines and hexoses after arterial injection. Am J Physiol 221:1629–1639

Olsson K (1969) Studies on central regulation of secretion of antidiuretic hormone (ADH) in the goat. Acta Physiol Scand 77:465–474

Olsson K (1970) Effects on water diuresis of infusions of transmitter substances into the third ventricle. Acta Physiol Scand 79:133–135

Olsson K (1972) On the importance of CSF Na^+ concentration in central control of fluid balance. Life Sci 11:397–402

Olsson K (1973) Further evidence for the importance of CSF Na^+ concentration in central control of fluid balance. Acta Physiol Scand 88:183–188

Olsson K, Kolmodin R (1974) Inhibition of antidiuretic hormone secretion by intracarotid infusions of slightly hypertonic non-electrolyte solutions. Acta Endocrinol 76:53–58

Olsson K, Larsson B, Liljekvist E (1976) Intracerebroventricular glycerol: A potent inhibitor of ADH-release and thirst. Acta Physiol Scand 98:470–477

Orloff J, Wagner HN, Davidson DG (1958) The effect of variations in solute secretion and vasopressin dosage on the excretion of water in the dog. J Clin Invest 37:458–464

Paintal AS (1973) Vagal sensory receptors and their reflex effects. Physiol Rev 53:159–227

Peck JW, Blass EM (1975) Localization of thirst and antidiuretic osmoreceptors by intracranial injection in rats. Am J Physiol 228:1501–1509

Phillips MI (1978) Angiotensin in the brain. Neuroendocrinology 25:354–377

Pickford M (1947) The action of acetylcholine in the supraoptic nucleus of the chloralosed dog. J Physiol (Lond) 106:264–270

Quillen EW, Cowley AW (1983) Influence of volume changes on osmolality-vasopressin relationships in conscious dogs. Am J Physiol 244:H73–H79

Ramsay DJ, Keil LC, Sharpe MC, Shinsako J (1978) Angiotensin infusion increases vasopressin, ACTH and II-hydroxycorticosteroid secretion. Am J Physiol 234:R66–R71

Reid IA, Brooks VL, Rudolph CD, Keil LC (1982) Analysis of the actions of angiotensin on the central nervous system of conscious dogs. Am J Physiol 243:R82–R91

Robertson GL (1977) The regulation of vasopressin function in health and disease. Rec Prog Horm Res 33:333–386

Robertson GL, Athar S (1976) The interaction of blood osmolality and blood volume in regulating plasma vasopressin in man. J Clin Endocrinol Metab 42:613–620

Robertson GL, Mahr EA, Athar S, Sinha T (1973) The development and clinical application of a new method for the radioimmunoassay of arginine vasopressin in human plasma. J Clin Invest 52:2340–2352

Robertson GL, Aycinena P, Zerbe RL (1982) Neurogenic disorders of osmoregulation. Am J Med 72:339–353

Robillard JE, Weitzman RE (1980) Developmental aspects of the fetal renal response to exogenous arginine vasopressin. Am J Physiol 238:F407–F414

Robillard JE, Weitzman RE, Fisher DA, Smith Jr FG (1979) The dynamics of vasopressin release and blood volume regulation during fetal hemorrhage in the lamb fetus. Pediatr Res 13:606–610

Rose JC, Meis PS, Morris M (1981) Ontogeny of endocrine (ACTH vasopressin cortisol) response to hypotension in lamb fetuses. Am J Physiol 240:E656–E661

Rundgren M, Fyhrquist F (1978) A study of permanent adipsia induced by medial forebrain lesions. Acta Physiol Scand 103:463–471

Rundgren M, McKinley MJ, Leksell LG, Andersson B (1979) Inhibition of thirst and apparent ADH release by intracerebroventricular ethacrynic acid. Acta Physiol Scand 105:123–125

Saavedra JM, Chevillard C (1982) Vasopressin-reversible increase in angiotensin-converting enzyme in specific hypothalamus nuclei of Brattleboro rats. Brain Res 246:157–160

Saito T, Yoshida S, Nakao K (1969) Release of antidiuretic hormone from neurohypophysis in response to hemorrhage and infusion of hypertonic saline in dogs. Endocrinology 85:72–78

Sawchenko PE, Friedman MI (1979) Sensory functions of the liver – a review. Am J Physiol 236:R5–R20

Schneider EG, Davis JO, Robb CA, Baumber JS, Johnson JA, Wright FS (1970) Lack of evidence for a hepatic osmoreceptor mechanism in conscious dogs. Am J Physiol 218:42–45

Schrier RW, Bichet DG (1981) Osmotic and nonosmotic control of vasopressin release and the pathogenesis of impaired water excretion in adrenal, thyroid and edematous disorders. J Lab Clin Med 98:1–15
Schrier RW, Goldberg JP (1980) The physiology of vasopressin release and the pathogenesis of impaired water excretion in adrenal, thyroid and edematous disorders. Yale J Biol Med 53:525–541
Schrier RW, Berl T, Herbottle JA (1973) Mechanism of effect of alpha adrenergic stimulation with norepinephrine on renal water excretion. J Clin Invest 53:502–511
Schrier RW, Berl T, Anderson RJ (1979) Osmotic and nonosmotic control of vasopressin release. Am J Physiol 236:F321–32
Schultz HD, Fater DC, Sundet WD, Geer PG, Goetz KL (1982) Reflexes elicited by acute stretch of atrial vs. pulmonary receptors in conscious dogs. Am J Physiol 242:H1065–H1076
Schwartz J, Reid IA (1981) Effect of vasopressin blockade on blood pressure regulation during hemorrhage in conscious dogs. Endocrinology 109:1778–1780
Segar WE, Moore WW (1968) The regulation of antidiuretic hormone release in man. I: Effects of change in position and ambient temperature on blood ADH levels. J Clin Invest 47:2143–2151
Share L (1965) Effects of carotid occlusion and left atrial distention on plasma vasopressin titer. Am J Physiol 208:219–223
Share L (1967) Role of peripheral receptors in the increased release of vasopressin in response to hemorrhage. Endocrinology 81:1140–1167
Share L (1968) Control of plasma ADH titer in hemorrhage: role of atrial and arterial receptors. Am J Physiol 215:1384–1389
Share L (1979) Interrelations between vasopressin and the renin-angiotensin system. Fed Proc 38:2267–2271
Share L, Levy MN (1962) Cardiovascular receptors and blood titer of antidiuretic hormone. Am J Physiol 203:425–428
Share L, Levy MN (1966) Carotid sinus pulse pressure, a determinant of plasma antidiuretic hormone concentration. Am J Physiol 211:721–724
Skowsky WR, Rosenbloom AA, Fisher DA (1974) Radioimmunoassay measurement of arginine vasopressin in serum: development and application. J Clin Endocrinol Metab 38:278–287
Sladek CD (1980) Osmotic control of vasopressin release: role of acetylcholine and angiotensin. In: Yoshida S, Share L, Yaki K (eds) Antidiuretic hormone. Japan Scientific Societies Press, Tokyo, pp 117–132
Sladek CD, Joynt RJ (1979) Cholinergic involvement in osmotic control of vasopressin release by the organ-cultured rat hypothalamo-neurohypophyseal system. Endocrinology 105:367–371
Sladek CD, Joynt RJ (1980) Role of angiotensin in the osmotic control of vasopressin release by the organ-cultured rat hypothalamo-neurohypophyseal system. Endocrinology 106:173–178
Sladek JR, McNeill TH, Khachaturian H, Zimmerman EA (1980) Chemical neuroanatomy of mono-amine interactions in the hypothalamic magnocellular system. In: Yoshida S, Share L, Yaki K (eds) Antidiuretic hormone. Japan Scientific Societies Press, Tokyo, pp 3–17
Sladek CD, Blair ML, Ramsay DJ (1982) Further studies on the role of angiotensin in the osmotic control of vasopressin release by the organ cultured rat hypothalamo-neurohypophyseal system. Endocrinology 111:599–607
Swaminathan S (1980) Osmoreceptors or sodium receptors: an investigation into ADH release in the rhesus monkey. J Physiol (Lond) 307:71–83
Szczepanska-Sadowska E (1974) Plasma ADH increase and thirst suppression elicited by preoptic heating in the dog. Am J Physiol 226:155–161
Thames MD, Schmid PG (1979) Cardiopulmonary receptors with vagal afferents tonically inhibit ADH release in the dog. Am J Physiol 237:H299–H304
Thrasher TN, Jones RG, Keil LC, Brown CJ, Ramsay DJ (1980a) Drinking and vasopressin release during ventricular infusions of hypertonic solutions. Am J Physiol 238:340–345

Thrasher TN, Brown CJ, Keil LC, Ramsay DJ (1980b) Thirst and vasopressin release in the dog: an osmoreceptor or sodium receptor mechanism. Am J Physiol 238:R333–339

Thrasher TN, Nistal-Herrera JF, Keil LC, Ramsay DJ (1981) Satiety and inhibition of vasopressin secretion after drinking in dehydrated dogs. Am J Physiol 240:E394–401

Thrasher TN, Keil LC, Ramsay DJ (1982a) Lesions of the organum vasculosum of the lamina terminalis (OVLT) attenuate osmotically induced drinking and vasopressin secretion in the dog. Endocrinology 110:1837–1839

Thraser TN, Keil LC, Ramsay DJ (1982b) Hemodynamic, hormonal and drinking response to reduced venous return in the dog. Am J Physiol 243:354–362

Tuomisto L, Eriksson L, Fyhrquist F (1980) Vasopressin release by histamine in the conscious goat. Eur J Pharmacol 107:334–341

Vandeputte-Van Messom G, Peeters G (1975) Effect of intraventricular administration of noradrenaline on water diuresis in goats. J Endocrinol 66:375–383

Vandeputte-Van Messom G, Peeters G (1976) Effect of intraventricular administration of isoprenaline on urinary function in the goat. Arch Int Pharmacodyn Ther 222:259–266

Vandeputte-Van Messom G, Peeters G (1979) Effect of intracerebroventricular administration of dopamine on urinary function in goats. Arch Int Pharmacodyn Ther 241:108–120

Van Wimersma Greidanus TB, Thody TJ, Verspaget H, deRotte GA, Goedemans HJH, Croiset G, Van Ree J (1979) Effects of morphine and β-endorphin on basal and elevated plasma levels of α-MSH and vasopressin. Life Sci 24:579–586

Verney EB (1947) The antidiuretic hormone and the factors which determine its release. Proc R Soc London Ser B 135:25–106

Vincent JD, Hayward JN (1970) Activity of single cells in the osmoreceptor-supraoptic nuclear complex in the hypothalamus of the waking rhesus monkey. Brain Res 23:105–108

Vincent JD, Arnauld E, Nicolescu-Catargi A (1972a) Osmoreceptors and neurosecretory cells in the supraoptic complex of the unanaesthetized monkey. Brain Res 45:278–281

Vincent JD, Arnauld E, Bioulac B (1972b) Activity of osmosensitive single cells in the hypothalamus of the behaving monkey during drinking. Brain Res 44:371–384

von den Velden R (1913) Die Nierenwirkung von Hypophysenextracten beim Menschen. Berlin Klin Wochenschr 50:2083–2086

Wade CE, Bie P, Keil LC, Ramsay DJ (1982a) Osmotic control of plasma vasopressin in the dog. Am J Physiol 243:E287–E292

Wade CE, Bie P, Keil LC, Ramsay DJ (1982b) Effect of hypertonic intracarotid infusions on plasma vasopressin concentration. Am J Physiol 243:E522–E526

Weiss CS, Almi CR (1975) Lateral preoptic and lateral hypothalamic units: in search of the osmoreceptors for thirst. Physiol Behav 15:713–722

Weitzman RE, Glatz TH, Fisher DA (1978a) The effect of hemorrhage and hypertonic saline upon plasma oxytocin and arginine vasopressin in conscious dogs. Endocrinology 103:2154–2160

Weitzman RE, Farnsworth L, Macphee R, Wang C, Bennett C (1978b) Effect of opposing osmolar and volume factors on plasma arginine vasopressin in man. Miner Electrolyte Metab 1:43–47

Weitzman RE, Fisher DA, Robillard J, Erenberg A, Kennedy R, Smith F (1978c) Arginine vasopressin response to an osmotic stimulus in the fetal sheep. Pediatr Res 12:35–38

Weitzman RE, Reviczky A, Oddie TH, Fisher DA (1980) Effect of osmolality on arginine vasopressin and renin release after hemorrhage. Am J Physiol 238:E62–E69

Wintour EM, Congiu M, Hardy KJ, Hennessy DP (1982) Regulation of urine osmolality in fetal sheep. Q J Exp Physiol 67:427–435

Wood CE, Keil LC, Rudolph AM (1982) Hormonal and hemodynamic responses to vena caval obstruction in fetal sheep. Am J Physiol 243:E278–E286

Woods JW, Bard P, Bleier R (1966) Functional capacity of the deafferented hypothalamus: water balance and responses to osmotic stimuli in the decerebrate cat and rat. J Neurophysiol 29:751–767

Yagi K, Kannan H, Sawaki Y (1980) Electrophysiology of antidiuretic hormone-secreting neurons: afferent neural pathways from cardiovascular receptors. In: Yoshida S, Share L, Yagi K (eds) Antidiuretic hormone. Japan Scientific Societies Press, Tokyo, pp 81–96

Yamaguchi K, Hama M, Sakaguchi T, Negoro H, Kamoi K (1980) Effects of intraventricular injection of Sar-Ala-angiotensin II on plasma vasopressin level increased by angiotensin II and by water deprivation in conscious rats. Acta Endocrinol 93:407–412

Yaron M, Bennett CM (1978) Mechanisms of impaired water excretion in acute right ventricular failure in conscious dogs. Circ Res 42:801–805

Zehr JE, Johnson JA, Moore WW (1969) Left atrial pressure, plasma osmolality and ADH levels in the unanaesthetized ewe. Am J Physiol 217:1672–1680

Vasopressin, Cardiovascular Regulation, and Hypertension

W. RASCHER, R. E. LANG, and TH. UNGER [1]

Contents

1 Vasopressin

1.1 Introduction

In 1895, Oliver and Schäfer reported that in anesthetized animals the injection of pituitary extracts was associated with a rise in blood pressure. Three years later, Howell (1898) demonstrated that the blood pressure-increasing material arose from the posterior lobe of the pituitary. This vasoactive substance from pituitary extracts was later called vasopressin.

A different biological activity of posterior pituitary extracts was observed independently by Farini (1913) and by Von den Velden (1913). These authors demonstrated that in patients suffering from disorders of the pituitary gland with symptoms of diabetes insipidus, administration of extracts from the posterior pituitary resulted in a marked decrease in urine output and a rise in urine osmolality. Careful studies on the isolated kidney confirmed the potent antidiuretic activity of these extracts (Starling and Verney 1924). Subsequently, two different sub-

1 Department of Pediatrics, Department of Pharmacology and German Institute for High Blood Pressure Research, University of Heidelberg, D-6900 Heidelberg

Current Topics in Neuroendocrinology, Vol. 4

stances were isolated from posterior pituitary extracts: vasopressin, which elicits an increase in blood pressure, and oxytocin, which causes uterine contractions (Kamm et al. 1928).

Oxytocin was the first biologically active peptide to be synthesized in the laboratory (Du Vigneaud et al. 1954 b). The synthesis was the result of intensive studies on the characterization of the biologically active principles from posterior pituitary extracts. The structure of vasopressin was detected in 1951 in Du Vigneaud's laboratory (Turner et al. 1951) and its synthesis was accomplished in 1954 (Du Vigneaud 1954 a).

On the basis of experimental data and results from human studies, it became generally accepted that the physiological role of vasopressin was solely related to its antidiuretic action. Cardiovascular effects of vasopressin were thought to be elicited only by high, "pharmacological" doses, many times those required for maximal antidiuretic activity. Therefore this hormone was called antidiuretic hormone, although "unfortunately", the name vasopressin persists as the official U.S.P. designation" (Brazeau 1975).

In the last few years, however, interest in the cardiovascular actions of vasopressin and its role in cardiovascular control and hypertension has been renewed. In particular, the experimental work of Möhring (1978) has produced evidence in favor of a role of vasopressin as a pressor hormone in hypertensive rats. The synthesis of specific antagonists of the vasoconstrictor activity by Manning, and its pharmacological characterization by Sawyer, have greatly stimulated the interest in vasopressin for cardiovascular regulation (Manning et al. 1981; Sawyer et al. 1981 a). The biological significance of vasopressin in the control of blood pressure and blood flow is still controversial and the available literature contains no definite evidence that vasopressin is an important pressor hormone (Möhring 1978; Johnston et al. 1981; Aisenbrey and Berl 1982; Liard 1982).

This review opens with a brief characterization of the biology of vasopressin and synthetic vasopressin analogues. The role of circulating vasopressin as a vasoactive peptide is then discussed in detail. Finally, extrahypothalamic vasopressin pathways and their interactions with blood pressure control mechanisms in the central nervous system are described.

1.2 Biochemistry and Physiology of Vasopressin

The nonapeptides arginine-vasopressin and oxytocin, together with their respective neurophysins, are synthesized in separate magnocellular neurons of the supraoptic and paraventricular nuclei of the hypothalamus (Walter 1975; Brownstein et al. 1980). The neurophysins and their respective hormones are produced as part of a single common precursor molecule, packaged into neurosecretory vesicles, and transported axonally to the nerve endings located in the neurohypophysis, where they are stored and secreted into the circulation (Sachs and Takabakate 1964; Sachs et al. 1969; Gainer et al. 1977; Brownstein et al. 1980). Recently, the complete amino acid sequences of the bovine vasopressin-neurophysin II precursor and of the oxytocin-neurophysin I precursor have been discovered by deduction from cDNA prepared from hypothalamic mRNA (Land et al. 1982,

1983, and this volume). Bovine neurophysin I and II showed complete homology between amino acids 10 and 74, which is about two-thirds of the whole neurophysin molecule.

Vasopressin is released into the circulation from the neurohypophysis by an exocytotic process (Dreifuss 1975) in response to osmotic and nonosmotic stimuli (Robertson 1977; Schrier et al. 1979). The major physiological stimulus for secretion of vasopressin is a rise of osmotic pressure in the extracellular fluid. The osmoreceptors are located in the anterior part of the hypothalamus, adjacent to the supraoptic and paraventricular nuclei (Robertson et al. 1982). Mechanisms for the release of vasopressin involve catecholamines (Bridges and Thorn 1970; Hoffman et al. 1977b; Miller et al. 1979; Kimura et al. 1981a, b), acetylcholine (Bridges and Thorn, 1970; Hoffman et al. 1977a; Sladek and Joynt 1979a, b), angiotensin II (Keil et al. 1975; Hoffman et al. 1977a), opioid peptides (Rossier et al. 1980; Ishikawa and Schrier 1982), prostaglandins (Vilhardt and Hedquist 1970; Yamamoto et al. 1976), and calcium (Thorn et al. 1980; Ishikawa et al. 1982).

Volume depletion is the second major stimulus for vasopressin release. During hypovolemia, vasopressin secretion is mediated via arterial baroreceptors, atrial stretch receptors, and cardiac receptors with vagal afferents (Share and Levy 1966; Share 1961, 1974; Gauer et al. 1970; Goetz et al. 1975; Kirchheim 1976; Schrier et al. 1979; Thames and Schmid 1979, 1981). Vasopressin concentrations in plasma in response to changes in osmolality or blood volume have been examined by a sensitive radioimmunoassay (Robertson et al. 1973). From these and a variety of other studies (Dunn et al. 1973; Skowsky et al. 1974; Weitzman and Fisher 1977; Hayward et al. 1976) it has become accepted that in the hydrated state, vasopressin plasma levels range between 1 and 4 pg/ml, and during dehydration, vasopressin levels rise in parallel with osmolality, to reach maximum antidiuresis at a plasma concentration between 10 and 20 pg/ml. Levels of vasopressin in severly volume-depleted subjects greatly exceeded those found in hyperosmolar subjects. Following volume depletion, vasopressin concentrations of more than 100 pg/ml plasma may act not only as an antidiuretic hormone but also as a pressor agent. The osmotic release of vasopressin is modulated by changes of blood volume (Dunn et al. 1973; Robertson and Athar 1976; Quillen and Cowley 1983). Contraction of blood volume is associated with an enhanced sensitivity of the osmoreceptor, whereas blood volume expansion resulted in a blunted osmotic release of vasopressin. Vasopressin acts physiologically as an antidiuretic hormone in the kidney and has an important role in the regulation of salt and water balance. Whereas aldosterone primarily regulates the whole body sodium content, vasopressin is responsible for water homoeostasis and regulation of sodium concentration in the extracellular fluid volume (Young et al. 1977). Its antidiuretic action is mediated by binding to a specific receptor located at the serosal or blood site of the collecting duct in the kidney. The hormone activates adenylcyclase with subsequent formation of cyclic AMP (Dousa and Valtin 1976). The sequence of events after the generation of cyclic AMP is not fully understood, but finally pores are induced in the collecting duct luminal membrane, which leads to a marked increase in water permeability (Gluck and Al-Awqati 1980). Using freeze-fracture electron microscopy, it has been shown

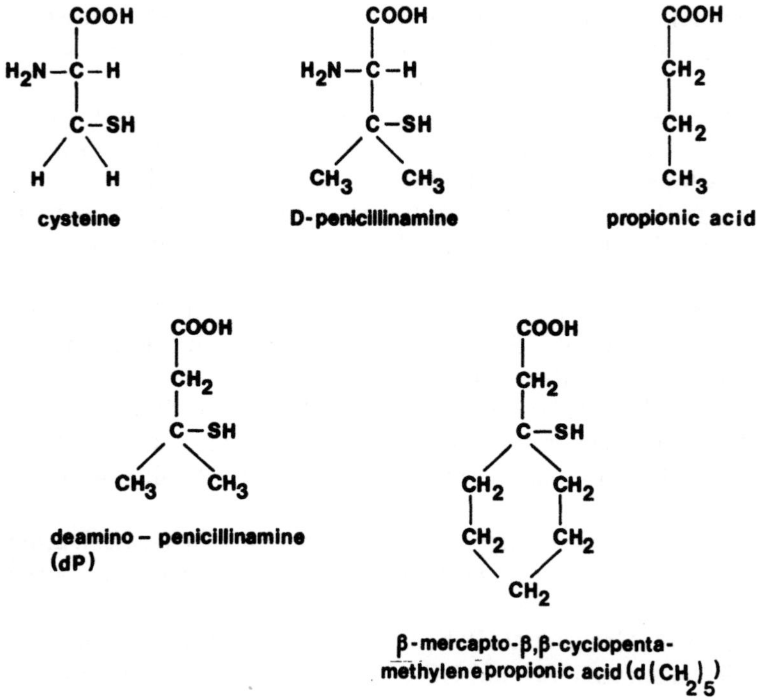

Fig. 1. Substituents of the vasopressin molecule at amino acid 1 (cysteine) by deamino-penicillinamine (dP) or by β-mercapto-β,β-cyclopentamethylenepropionic acid [$d(CH_2)_5$]. For comparison the structures of cysteine, penicillamine, and propionic acid are given

that intramembranous particle clusters on the collecting duct luminal membrane represent the specific structural change related to the action of vasopressin (Harmanci et al. 1980).

The characterization of oxytocin and arginine-vasopressin in the early 1950s stimulated not only the fields of synthetic peptide chemistry and peptide endocrinology in general, but also the investigation of structure-activity relationships of neurohypophyseal hormones, and particularly the search for structural analogues of these hormones with more specific actions.

Modifications of the vasopressin and oxytocin molecules have yielded a variety of structural analogues with more selective oxytocic, antidiuretic, or vasoconstrctior activities (Berde and Boissonnas 1968; Sawyer and Manning 1973; Manning et al. 1977a). More recently, the successful synthesis of specific and potent antagonists of the vasoconstrictor activity of vasopressin have been reported (Manning et al. 1981; Sawyer et al. 1981a). These compounds have greatly stimulated research on the role of vasopressin in cardiovascular regulation. In addition, Manning successfully synthesized antagonists of the antidiuretic activity of vasopressin (Manning et al. 1982a, b; Sawyer et al. 1981b). Various vasopressin analogues are discussed here briefly.

The available antagonists of the vasoconstrictor response of vasopressin (pressor antagonists) are essentially based on modifications of vasopressin itself or ⟨1-deamino, 4-valine, 8-D-arginine⟩ vasopressin (Manning et al. 1973). The

Table 1. Arginine-vasopressin (AVP) and antagonists of its vasoconstrictor activity (vascular antagonists). (Adapted from Manning et al. 1981)

Vasopressin analogue	Change in position	Introduced substituent	Antivasopressor activity (pA_2)	Antidiuretic activity (U/mg)
$d(CH_2)_5AVP$	1	$d(CH_2)_5$	8.35 ± 0.09	0.033 ± 0.005
dPTyr(Me)AVP	1	dP	7.96 ± 0.05	3.5 ± 0.5
	2	Tyr(Me)		
$d(CH_2)_5Tyr(Me)AVP$	1	$d(CH_2)_5$	8.62 ± 0.03	0.31 ± 0.07
	2	Tyr(Me)		
dPVDAVP	1	dP	7.82 ± 0.05	123 ± 22
	4	L-valine		
	8	D-arginine		
$d(CH_2)_5VDAVP$	1	$d(CH_2)_5$	7.68 ± 0.05	0.10 ± 0.02
	4	L-valine		
	8	D-arginine		

1 2 3 4 5 6 7 8 9
Cys-Tyr-Phe-Gln-Asn-Cys-Pro-Arg-GlyNH_2 AVP

$d(CH_2)_5$, 1-β-mercapto-β,β-cyclopentamethylenepropionic acid; dP, deamino-penicillamine; Tyr(Me), methyltyrosine

following modifications were made alone or in combination (Fig. 1): (1) Substitution with 1-D-penicillamine (dP) (Schulz and Du Vigneaud 1966); (2) Substitution with 1-β-mercapto-β,β-pentamethylenepropionic acid ($d(CH_2)_5$) (Nestor et al. 1975); (3) Methyl substitution of tyrosine [Tyr(Me)].

Some potent vascular vasopressin antagonists that have successfully been used as pharmacological tools to study the role of vasopressin in cardiovascular regulation are: ⟨1-Deamino-penicillamine-4-valine⟩ – 8-D-arginine-vasopressin (dPVDAVP) (Manning et al. 1977b), ⟨1-deamino-penicillamine, 2-(O-methyl) tyrosine⟩ arginine-vasopressin (dPTyr(Me)AVP) (Bankowski et al. 1978), and ⟨1-(β-mercapto-β,β-cyclopentamethylenepropionic acid)⟩, 4-valine, 8-D-arginine-vasopressin ($d(CH_2)_5$ VDAVP) (Lowbridge et al. 1978), as well as ⟨1-(β-mercapto-β,β-cyclopentamethylenepropionic acid), 2-(O-methyl) tyrosine)⟩ arginine-vasopressin ($d(CH_2)_5$ Tyr(Me)AVP) and ⟨1-(β-mercapto-β,β-cyclopentamethylenepropionic acid)⟩ arginine-vasopressin ($d(CH_2)_5$ AVP) (Kruszynski et al. 1980), (Table 1).

Substitution of the vasopressin molecule in position 2 by phenylalanine, position 3 by isoleucine, or position 8 by ornithine, resulted in decreased antidiuretic activity and enhanced vasopressor selectivity (pressor agonists). These substitutions have additive effects, and (2-phenylalanine, 3-isoleucine, 8-ornithine)-vasotocin is the most specific vasopressor analogue known (Berde et al. 1964). Selective vasopressor analogues are useful as local hemostatic agents and for the control of upper gastrointestinal bleeding, particular from esophageal varices.

Deamino-8-D-arginine vasopressin (DDAVP) is an analogue of vasopressin with potent and prolonged antidiuretic activity but without significant cardiovascular effects (antidiuretic agonist) (Vávra et al. 1968; Sawyer et al. 1974; Johnson et al. 1979), and is therefore now used as a drug of choice in the treatment of vasopressin-deficient diabetes insipidus.

Recently, analogues of vasopressin that antagonize the antidiuretic response have been reported (Manning et al. 1981, 1982a, b; Manning und Sawyer 1982; Sawyer et al. 1981b). The 2-O-ethylty rosine analogue of ⟨1-(β-mercapto-β,β-cyclopentamethylene propionic acid), 4-valine⟩ arginine vasopressin $d(CH_2)_5$ VAVP) appears to be the most potent anti-antidiuretic peptide tested so far (Sawyer et al. 1981b). The antagonists are probably useful agents for the treatment of hyponatremia and water retention secondary to inappropriate secretion of antidiuretic hormone.

2 Circulatory Effects of Vasopressin

2.1 Vascular and Hemodynamic Effects of Vasopressin

There are numerous studies demonstrating a direct vasoconstrictor activity of vasopressin preparations from posterior pituitary extracts, as well as of synthetic vasopressin (Nakano 1974; Monos et al. 1978; Richardson and Withrington 1978). Vasopressin in doses as low as those physiologically occurring in plasma (10^{-12} mol/l) constricts arterioles of the mesenteric vasculature (Altura 1973; Altura and Altura 1977). The vascular vasopressin receptor has recently been characterized in aortic smooth muscle cells in culture (Penit et al. 1983). Regional vascular beds possess different sensitivity to vasopressin. Iliac and mesenteric vasculature responded very sensitively to synthetic lysine-vasopressin, whereas this agent has only minimal effects on renal blood flow, (Heyndrickx et al. 1976). A similar redistribution of blood flow, to the kidney from the muscles and the gastrointestinal tract, caused by vasopressin (Pitressin), has also been demonstrated in anesthetized dogs (Schmid et al. 1974). From the study of Heyndrickx et al. (1976) it was suggested that the peripheral distribution of vascular vasopressin receptors was similar to that of alpha-adrenergic receptors. Studies in guinea pig mesenteric vessels suggest that vasopressin acts on small muscular arteries only, whereas norepinephrine acts on all mesenteric vessels, including the portal vein (Karashima 1981). Evidence has been presented that regional vascular beds may possess vasopressin receptors that differ in their relative sensitivity to vasopressin analogues (Altura 1975).

An interaction between vasopressin and catecholamines was demonstrated as early as 1912 by Kepinow at the Department of Pharmacology, University of Heidelberg. This author observed that a pituitary extract (hypophysin) augmented the blood pressure rise which occurred after administration of epinephrine in vivo and in vitro (Kepinow 1912). The fact that small doses of vasopressin enhance vasoconstriction induced by catecholamines is now well established (Chenoweth et al. 1958; Bartelstone and Nasmyth 1965; Erker and Chan 1977).

Vasopressin also interacts with angiotensin II (McNeill et al. 1970). McNeill (1974) observed that vasopressin and angiotensin are the most important deter-

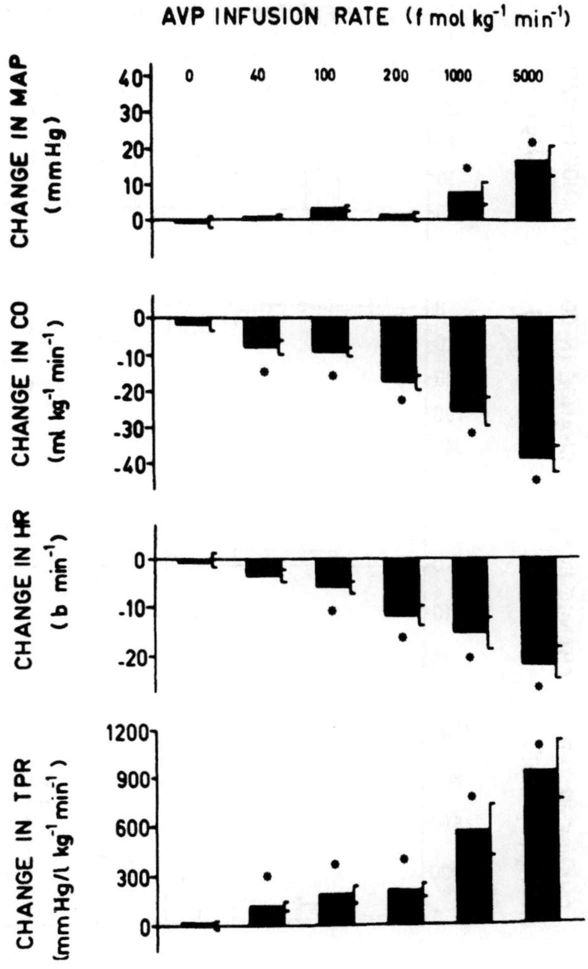

Fig. 2. Changes in mean arterial pressure (MAP), cardiac output (CO), heart rate (HR), and total peripheral resistance (TPR) induced by infusion of various doses of arginine-vasopressin in conscious dogs. *, significantly different compared to saline infusion (0). (Liard 1982)

minants of mesenteric vasoconstriction after hemorrhage or hypovolemia induced by diuretic agents. When the vasopressin-induced vasoconstriction was abolished by hypophysectomy, mesenteric resistance was maintained by an enhanced activity of the renin-angiotensin system (McNeill et al. 1976; Pang et al. 1979).

In rabbits with renal artery stenosis, the enhanced pressor responsiveness to vasopressin which occurs 3 days after clamping the kidney can be abolished by infusion of (Sar^1, Ile^8)angiotensin II, a competitive angiotensin II antagonist (Johnson et al. 1981).

Hatzinikolaou et al. (1980, 1981), when investigating the cause of blood pressure rise after infusion of hypertonic saline, found that in anephric rats vasopressin was responsible for about half of the observed blood pressure increment. In these experiments vasopressin-specific effects were demonstrated by using vasopressin-deficient diabetes insipidus rats and specific vascular vasopressin antagonists. In normal rats after catecholamine depletion, vasopressin and angiotensin

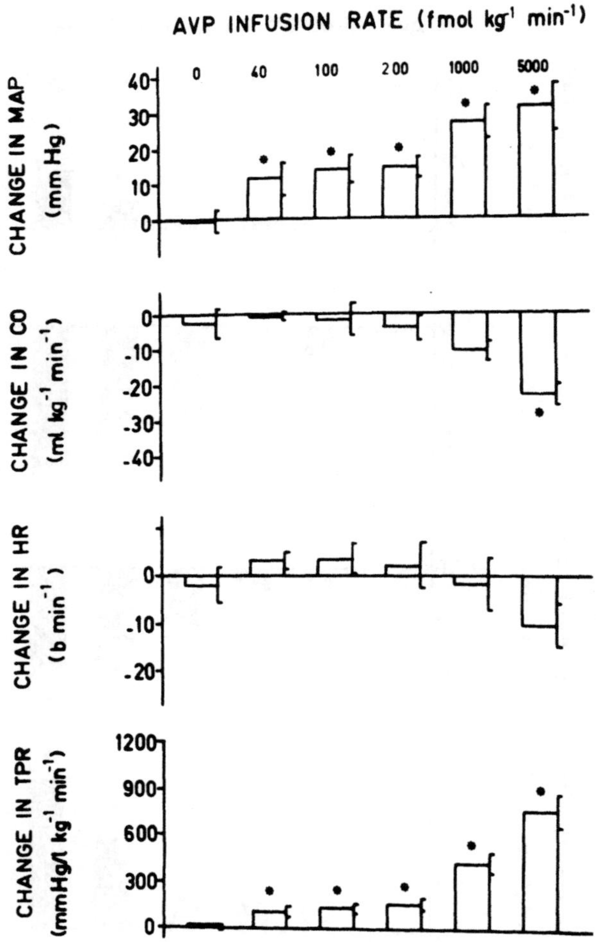

Fig. 3. Changes in mean arterial pressure (MAP), cardiac output (CO), heart rate (HR), and total peripheral resistance (TPR) induced by infusion of various does of arginine-vasopressin in conscious baroreceptor-denervated dogs. *, significantly different compared to saline infusion (0). (Liard 1982)

play an important role in maintaining arterial blood pressure (Gavras et al. 1982).

Hemodynamic effects of vasopressin have recently been studied in conscious animals. Continuous infusion of vasopressin in low doses resulted in a transient rise in blood pressure which was probably mediated by hypervolemia (Smith et al. 1979). Szczepanska-Sadowska (1973) reported that moderate doses of vasopressin (Pitressin) infused in conscious dogs increased mean arterial blood pressure, central venous pressure, systemic peripheral resistance, and plasma volume, whereas cardiac output and heart rate fell. In these experiments, the average antidiuretic activity in plasma that exerted these cardiovascular effects was 22 μU/ml (=57 pg/ml) when measured by bioassay technique. The hemodynamic consequences of more physiological levels of vasopressin were further studied in conscious dogs with and without sino-aortic baroreceptor deafferentation by Cowley et al. 1974. These authors observed an average increase in mean arterial pressure of 5 mmHg in intact dogs during vasopressin infusion (1.6 mU/kg/min), whereas after baroreceptor deafferentation blood pressure rose by 33 mmHg. Montani et al. (1980) were able to demonstrate for the first time in a conscious dog, that ar-

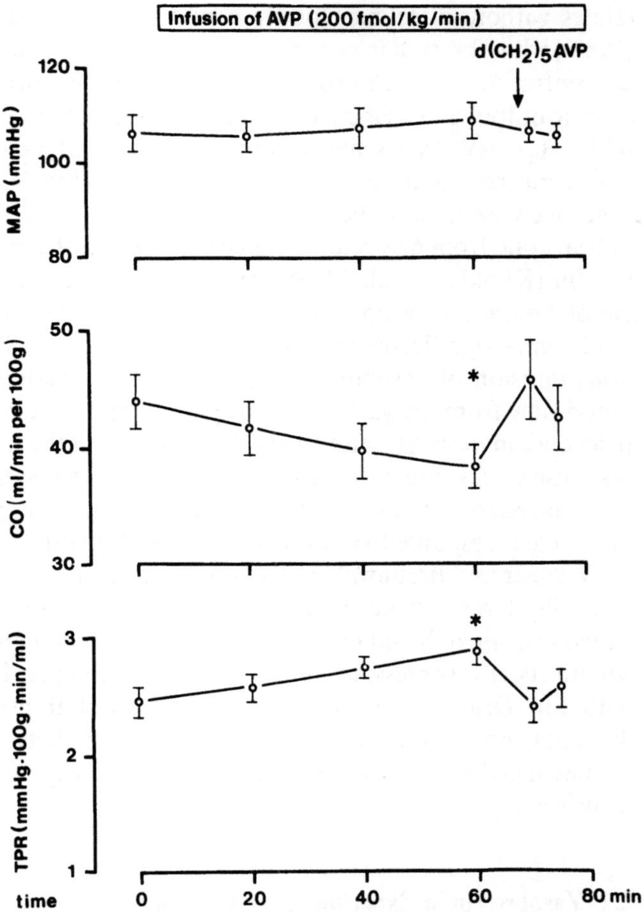

Fig. 4. Effects of intravenous infusion of arginine-vasopressin (AVP) in a dose of 200 fmol/kg/min and of a superimposed intravenous injection of the vascular vasopressin antagonist $d(CH_2)_5$ AVP on mean arterial pressure (MAP), cardiac output (CO), and total peripheral resistance (TPR) in conscious rats. *, significantly different compared to preinfusion values (0). Data published in Rascher et al. (1983)

ginine-vasopressin in physiological concentrations had significant hemodynamic effects. Increases in plasma arginine-vasopressin of 2–20 pg/ml after infusion of these hormones, or release of arginine-vasopressin by increasing plasma osmolality following hypertonic saline infusion, did not affect arterial blood pressure. However, cardiac output fell and total peripheral resistance rose in dose-dependent fashion (Fig. 2). After sino-aortic baroreceptor deafferentation, small amounts of vasopressin (40 fmol/kg per min) which doubled plasma vasopressin concentration, resulted in an increase of blood pressure and total peripheral resistance. Under these circumstances cardiac output did not change (Fig. 3). These observations suggested that vasopressin in physiological plasma concentrations causes vasoconstriction but does not influence arterial blood pressure, since blood pressure effects are effectively buffered by the baroreceptor reflex mechanism via reduction of cardiac output. The fact that vasopressin has hemodynamic effects without affecting blood pressure has been confirmed by Rascher et al. (1983), who observed in conscious rats that a two- to threefold rise in vasopressin concentration (through exogenous infusion of the hormone) resulted in a fall in cardiac output and an increase in total peripheral resistance. After administration of $d(CH_2)_5$ AVP, a vascular vasopressin antagonist, cardiac output rose and total peripheral resistance fell to preinfusion values (Fig. 4). In normal volunteers, a slight increase in diastolic blood pressure from 72 ± 3 to 78 ± 2 mmHg and a fall in heart rate from 67 ± 5 to 63 ± 5 beats/min was observed after infusion of vasopressin (Khokhar et al. 1980). Systolic blood pressure did not change. The increased release of vasopressin in newborn infants has been suggested to be of hemodynamic significance (Pohjavuori and Fyhrquist 1980). Furthermore, intravenous infusion of vasopressin in pregnant sheep resulted in a redistribution of blood flow from the gastrointestinal and peripheral circulation to the umbilical-placental, myocardial, and cerebral circulation in the fetus. Since cardiovascular response to vasopressin infusion was similar to that seen during fetal hypoxia, it was suggested that vasopressin release my play an important role in fetal cardiovascular response to stress (Iwamoto et al. 1979).

Wagner and Braunwald (1956) observed in patients with orthostatic hypotension (Shy-Drager syndrome) that infusion of very small doses of vasopressin resulted in a rise in blood pressure. Recently, a similar enhanced sensitivity to pressor effects of vasopressin was observed by Möhring et al. (1980 a). In two patients with Shy-Drager syndrome, increases in arterial pressure of about 20 and 40 mmHg occurred with plasma vasopressin levels between 5 and 10 pg/ml. These studies also demonstrated the potent pressor activity of vasopressin in autonomic insufficiency.

2.2 Vasopressin in Experimental Hypertension

Page and Sweet reported in 1937 that hypophysectomy decreased blood pressure in dogs made hypertensive by renal artery stenosis. Although these studies may demonstrate a role of vasopressin in the pathogenesis of hypertension, vasopressin as a pressor hormone was neglected for many years. The work of Möhring (1978) renewed interest in the role of vasopressin in the pathogenesis of hypertension. He originally suggested that in different forms of experimental hypertension of rats (renal, DOCA, and genetic), vasopressin had a role as pressor hormone during the established phase of hypertension. Particularly, vasopressin was suggested to be an important pressor hormone in the malignant phase of high blood pressure (Möhring 1978). The conclusion was based on the finding that plasma vasopressin was above normal in experimental models of hypertension, and that blood pressure was transiently lowered in these experimental models by intravenous injection of a specific vasopressin antiserum. Since plasma levels of vasopressin in hypertensive animals are only slightly above normal and far below pressor levels achieved by exogenous infusion of vasopressin, a specific sensitization to the pressor effects of this hormone was postulated (Möhring 1978). This could be due either to a reduced gain of the baroreceptor reflex or to increased sensitivity of the vascular smooth muscle (Möhring 1978; Möhring et al. 1979).

2.2.1 DOCA Hypertension

In 1960 Friedman et al. demonstrated that administration of vasopressin (Pitressin) accelerated the onset of hypertension induced by overdosage of the mineralocorticoid hormone desoxycorticosterone acetate (DOCA). Furthermore, surgical ablation of the median eminence, a procedure inducing central diabetes insipidus, prevented the development of DOCA hypertension (Friedman et al. 1960). Subsequently Möhring et al. (1976, 1977) reported increased plasma concentrations of vasopressin in DOCA-hypertensive rats, which has been confirmed by other authors (Crofton et al. 1980; Rascher et al. 1981 a, 1983; Morton et al. 1982). The fact that injection of a specific vasopressin antiserum transiently lowered mean blood pressure led to the hypothesis that vasopressin had a role as a pressor hormone in this form of experimental hypertension (Möhring et al. 1976, 1977). Studies reexamining the pressor role of this hormone in DOCA-hypertensive rats by using specific antagonists of the vasopressor response to arginine-vasopressin instead of antibodies yielded contradictory results. Crofton et al. (1979) observed a significant reduction in blood pressure following intravenous injection of two vascular vasopressin antagonists [dPVDAVP and $d(CH_2)_5$ VDAVP] after several weeks of treatment with DOCA. In contrast, Rabito et al. (1981), using the same vasopressin antagonists, did not find a fall in blood pressure in DOCA-hypertensive rats. These authors have criticized the use of antibodies by Möhring et al. (1977) and suggested that the antibodies may have lowered blood pressure by a non-specific antigen-antibody reaction and not by blockade of vasopressin. Unfortunately, Rabito et al. (1981) did not measure vasopressin in their rats, and they injected vasopressin, norepinephrine, and angiotensin II prior to injection of the vasopressin antagonist. Rascher et al. (1981 a, 1983) also reported that two vascular vasopressin antagonists [$d(CH_2)_5$ VDAVP and $d(CH_2)_5$ AVP] had no effect on mean arterial blood pressure in rats treated for 4 weeks with DOCA. However, a significant fall in total peripheral resistance was observed, as well as an increase in cardiac output. After sino-aortic deafferentation, $d(CH_2)_5$ AVP decreased arterial blood pressure and total peripheral resistance, since cardiac output did not change in these rats (Fig. 5). This suggested that the baroreceptor reflex buffered the vascular effects of vasopressin by reducing cardiac output. The fact that the function of the baroreceptor reflex mechanism interferes with the vasoconstrictor activity of vasopressin may explain why some investigators did and others did not observe a fall in blood pressure after administration of the same vascular vasopressin antagonist. The reduction of blood pressure observed by Crofton et al. (1979) after several weeks of treatment with DOCA may have been the consequence of an already impaired baroreceptor reflex in the established phase of DOCA hypertension, and thus a diminished cardiac response. This explanation is corroborated by the fact that a progressive alteration of the baroreceptor reflex has been reported in DOCA-hypertensive rats (Matsugushi and Schmid 1982). Such a resetting of the reflex might also explain the enhanced systemic pressor response to vasopressin in rats treated for 6–8 weeks with DOCA (Crofton et al. 1980).

Wisgerhof et al. (1977) found that Brattleboro rats with hypothalamic diabetes insipidus do not develop hypertension after treatment with DOCA. Vaso-

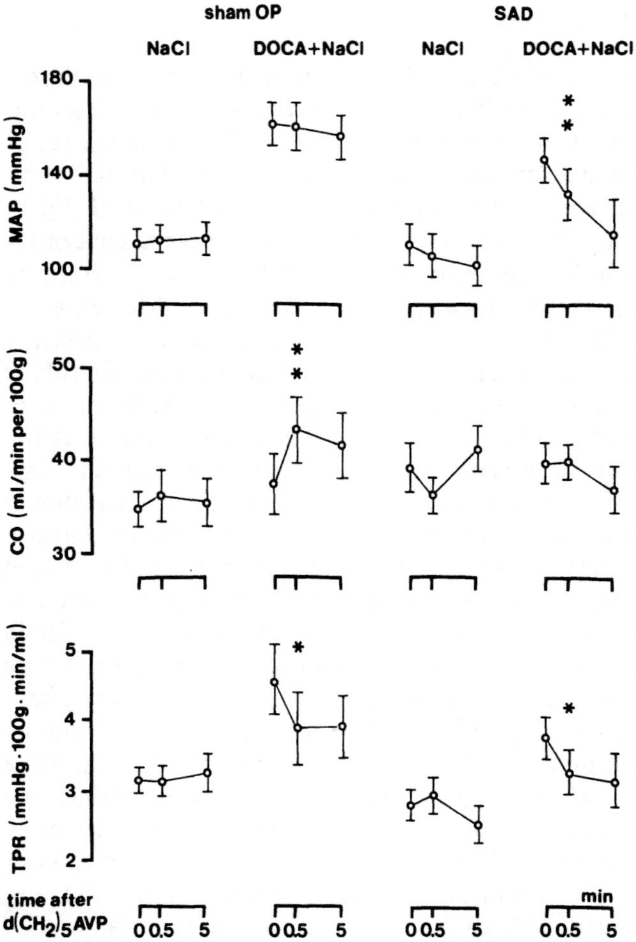

Fig. 5. Effect of an intravenous injection of the vascular vasopressin antagonist $d(CH_2)_5AVP$ (5μg/kg) on mean arterial pressure (MAP), cardiac output (CO) and total peripheral resistance (TPR) in conscious DOCA-hypertensive and normotensive control rats without (*sham* OP) and with (*SAD*) sino-aortic deafferentation. Rats received 1% NaCl solution to drink. *, $p<0.025$; **, $p<0.01$ as compared with the control value (time 0). Rascher et al. 1983)

pressin (Pitressin) treatment restored the hypertensive effects of DOCA in these rats. Berecek et al. (1982) confirmed these findings and demonstrated that hypertension in vasopressin-treated diabetes insipidus rats was associated with an induction of increased renal vascular reactivity. Saito and Yajima (1982) reported that infusion of DDAVP, a potent antidiuretic antagonist of vasopressin with markedly reduced vasoconstrictor activity, also restored the hypertensive response to DOCA in Brattleboro rats. In DOCA-treated diabetes insipidus rats, arginine vasopressin in a dose comparable in antidiuretic activity to the DDAVP dose used led to a more marked rise in blood pressure than DDAVP, and therefore Saito and Yajima concluded that both the antidiuretic and the pressor activity of vasopressin were essential for the establishment of DOCA hypertension.

The available data speak in favor of a possible role for vasopressin in DOCA hypertension of rats. However, the exact mechanism of involvement of vasopressin remains unresolved.

2.2.2 Genetic Hypertension

Elevated plasma concentrations of arginine-vasopressin have been reported in spontaneously hypertensive rats of the Okamoto strain (SHR) (Crofton et al. 1978) and in stroke-prone SHR (SHRSP) during the established phase of hypertension (Möhring et al. 1978b, 1979) when compared with normotensive Wistar-Kyoto rats. However, Rascher et al. (1981b) reported in the same model of SHRSP used by Möhring that vasopressin concentrations are above normal only if the rats are more than 24 weeks of age. In 6-, 9-, and 12-week-old rats, plasma vasopressin levels were lower than in age-matched normotensive Wistar-Kyoto rats. Therefore, in contrast to DOCA hypertension, plasma vasopressin appears to play no role in the development of hypertension in these rats. Crofton et al. (1978), who also reported increased urinary excretion and pituitary content of vasopressin in SHR, did not observe a significant fall in blood pressure after injection of vascular vasopressin antagonists in these rats, and they expressed doubt concerning the involvement of vasopressin in the pathogenesis of genetic hypertension in rats. Möhring et al. (1979), however, concluded that elevated plasma vasopressin levels played an important role in the maintenance of high blood pressure in SHR 22–28 weeks of age. These authors found a positive correlation between blood pressure and plasma vasopressin concentration, and observed a fall in blood pressure after injection of a specific vasopressin antiserum. In a subsequent study they observed a marked reduction in cardiovascular reflex acitvity during infusion of vasopressin in SHR (Möhring et al. 1981). In sharp contrast to this, Rascher et al. (1982d) presented evidence against a role of circulating vasopressin in the pathogenesis of spontaneous hypertension in SHRSP. These authors did not observe a fall in blood pressure or peripheral resistence in 40-week-old SHRSP after administration of the vascular vasopressin antagonist $d(CH_2)_5$ AVP.

Definite evidence that vasopressin is not necessary for the development of hypertension in SHRSP has been recently presented by Ganten et al. (1983). These authors reported the successful development of a new strain of SHR with hypothalamic diabetes insipidus. By crossbreeding SHRSP with Brattleboro rats homozygous for diabetes insipidus, the diabetes insipidus gene was successfully introduced into the SHRSP strain. These SHRDI rats all had typical signs of hypothalamic diabetes insipidus, with no detectable concentrations of vasopressin in the plasma, hypothalamus, and pituitary gland. Despite lack of vasopressin, mean arterial pressure increased to 184 ± 9.7 mmHg as compared to 197 ± 5.2 mmHg in original SHRSP rats (Fig. 6). It can be concluded with certainty from these results that vasopressin is not essential for the development and maintenance of spontaneous hypertension of rats.

Matsugushi et al. (1981) reported increased levels of vasopressin in salt-sensitive compared to salt-resistant genetically hypertensive rats of the DAHL strain. Injection of the vascular vasopressin antagonist $d(CH_2)_5$ VDAVP did not sig-

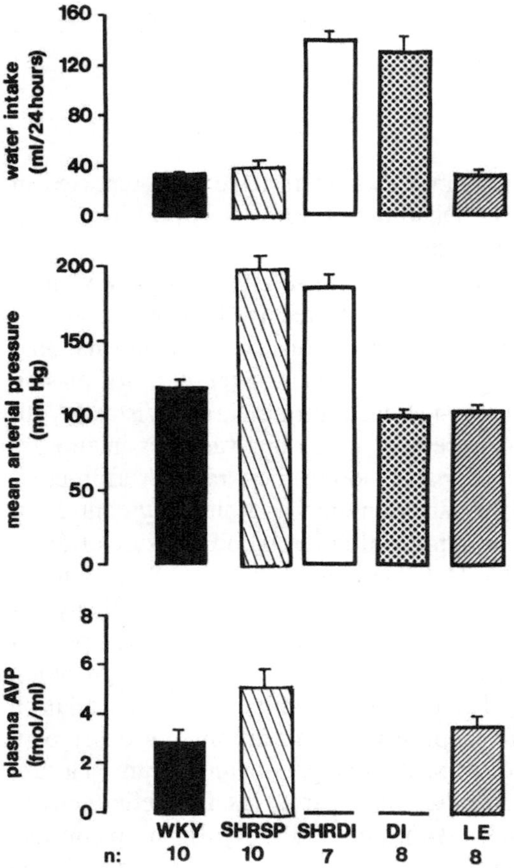

Fig. 6. Water intake, mean arterial pressure, and plasma arginine-vasopressin (AVP) concentration in spontaneously hypertensive diabetes insipidus rats (*SHRDI*), spontaneously hypertensive rats of the stroke-prone substrain (*SHRSP*), normotensive Wistar-Kyoto rats (*WKY*), diabetes insipidus rats (*DI*), and Long-Evans control rats (*LE*). (Ganten et al. 1983)

nificantly lower arterial blood pressure in either strain (Matsugushi et al. 1981). Share et al. (1982) confirmed these findings, and furthermore observed that the vascular vasopressin antagonist $d(CH_2)_5$ Tyr(Me) AVP lowered blood pressure in salt-sensitive DAHL rats on a high sodium diet, when these were pretreated with the converting enzyme inhibitor captopril (Fig. 7). In genetically hypertensive rats of the New Zealand strain, urinary vasopressin excretion and plasma vasopressin levels were unchanged (Crofton et al. 1981). However, pressor response to vasopressin was markedly increased in this form of genetic hypertension of rats.

2.2.3 Renal Hypertension

In rats made hypertensive by 70% reduction of the renal mass and a 1% saline solution as drinking fluid, plasma vasopressin levels and urinary excretion of vasopressin were elevated (Lee-Know et al. 1981). These authors observed that the vascular vasopressin antagonist dPVDAVP lowered blood pressure in hypertensive rats only by 9 mmHg. They concluded that vasopressin was not a major

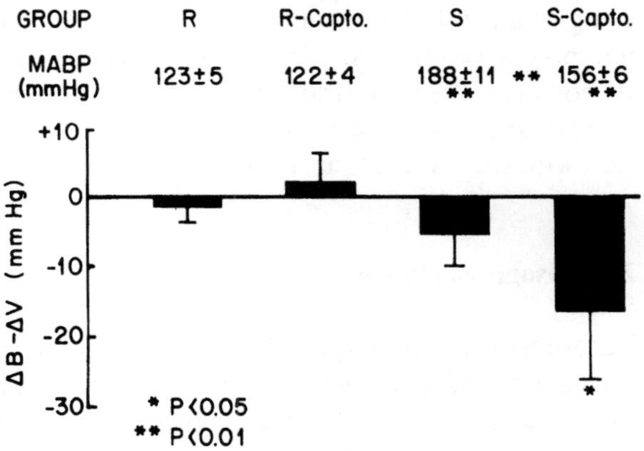

Fig. 7. Changes in mean arterial blood pressure (MABP) following intravenous injection of the vascular vasopressin antagonist $d(CH_2)_5$ Tyr(Me)AVP in DAHL resistant (*R*) and sensitive rats. R-Capto and S-Capto rats were pretreated with captopril. The bars indicate the differences between the changes in MABP in response to the blocker (ΔB) and to the 0.9% saline vehicle (ΔV). (Share et al. 1982)

pressor agent in partial nephrectomy–salt hypertension. Di Pette et al. (1982) reported that in rats with 85% of the renal mass surgically removed a 24-h infusion of 0.9% saline solution resulted in a marked acute rise of arterial blood pressure by 29 mmHg and an average elevation of plasma vasopressin levels to 42 pg/ml. Injection of the vascular vasopressin antagonist $d(CH_2)_5$ Tyr(Me)AVP resulted in a fall in blood pressure of 20 mmHg. The authors concluded that vasopressin plays an important role in blood pressure elevation induced acutely by salt excess in rats with renal insufficiency. After reduction of renal mass by partial nephrectomy, infusion of vasopressin and hypotonic saline caused a more chronic elevation of blood pressure as a consequence of positive water balance (Manning et al. 1979). Lohmeier et al. (1981) demonstrated that vasopressin had pronounced acute effects on blood pressure, but that long-term changes were minimal or nonexistent.

In rats made hypertensive by unilateral renal artery stenosis, with the contralateral kidney untouched (two kidney–one clip hypertension), plasma vasopressin levels are elevated and blood pressure decreased transiently toward normal after administration of a vasopressin antiserum, especially in the "malignant phase" of this form of hypertension (Möhring et al. 1978 a). However, Rabito et al. (1981) did not find a fall in blood pressure after administration of a vascular vasopressin antagonist in renal hypertensive rats. Furthermore, Pullan et al. (1978, 1980), studying dogs with renal artery stenosis of the remaining kidney after unilateral nephrectomy (one kidney–one clip hypertension) disputed a pathogenic role of vasopressin in this form of hypertension. These authors observed a marked rise of mean arterial blood pressure at a time when plasma vasopressin levels had only doubled and they concluded that vasopressin had no significant direct vasoconstrictor role in this form of experimental renal hypertension. Furthermore, it has been demonstrated that Brattleboro rats with hypothalamic diabetes insipidus develop two kidney–one clip hypertension despite an inability to synthesize vasopressin (Woods and Johnston 1982). The development of one kidney–one clip hypertension was similar in diabetes insipidus rats and Long-Evans control rats. However, the degree of hypertension was significantly less in diabetes insipidus rats. Treatment with DDAVP, the synthetic vasopressin antidiuretic agonist with minimal vasoconstrictor activity, elevated blood pressure to similar levels as in

Long-Evans rats. The same laboratory presented evidence that vasopressin does not play a primary role in the pathogenesis of malignant renal hypertension (Wood et al. 1983). In Brattleboro rats with a total defect in the synthesis of vasopressin, aortic ligature between the renal arteries produced malignant hypertension with the same frequency and severity as in Long-Evans control rats.

2.3 Vasopressin in Human Hypertension

Vasopressin as a pathogenic factor in human hypertension was first suggested by Ellis and Grollman (1949), when they observed enhanced antidiuretic activity in the urine of patients with severe hypertension. Increased renal excretion of vasopressin was also reported in young men with mild essential benign hypertension (Khokhar and Slater 1976). Plasma vasopressin measurements in patients with essential hypertension revealed high vasopressin levels during the malignant phase, whereas in the benign stage vasopressin levels tend to be lower than in normotensive controls (Padfield et al. 1976). When vasopressin was infused in five volunteers to produce concentrations five times higher than observed in patients with malignant hypertension, blood pressure did not change (Padfield et al. 1976). These authors also examined the relationship between plasma vasopressin and blood pressure in 29 patients with malignant hypertension and 28 patients with the syndrome of inappropriate antidiuretic hormone secretion (Padfield et al. 1981). They did not find a correlation between the elevated plasma vasopressin levels in malignant hypertension (13 ± 3 pg/ml) and arterial blood pressure. In patients with inappropriate antidiuretic hormone secretion blood pressure was normal and did not correlate with the rise in plasma vasopressin (39 ± 7 pg/ml).

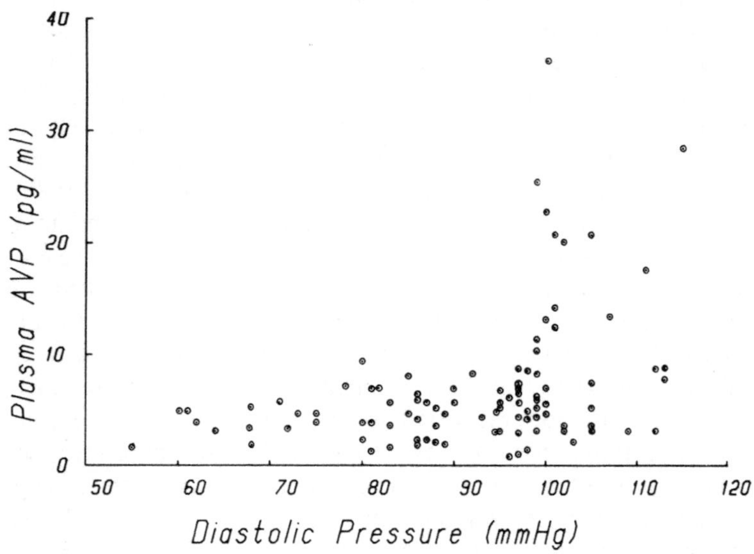

Fig. 8. Relationship between plasma arginine-vasopressin (AVP) and diastolic blood pressure in patients with essential hypertension. (Cowley et al. 1981)

Cowley et al. (1981) noted elevated plasma vasopressin levels in patients with essential hypertension, particularly when diastolic pressure exceeded 95 mmHg (Fig. 8). In hypertensive subjects older than 50 years, high sodium intake appeared to be correlated with the rise in plasma vasopressin. The role of vasopressin in human essential hypertension is not yet clear. The available results rather argue against a significant direct influence of plasma vasopressin on arterial blood pressure.

2.4 Vasopressin in Dehydration and Hemorrhage

During dehydration there is a small rise in plasma vasopressin levels in man (Robertson et al. 1973) and in animals (Dunn et al. 1973). Robertson et al. demonstrated a close direct relationship between plasma vasopressin concentration and plasma osmolality, which has been confirmed by many authors. It was generally accepted that a small increase in plasma vasopressin up to 20 pg/ml after dehydration is not associated with hemodynamic effects. However, two recent reports using specific vascular vasopressin antagonists raised the possibility that vasopressin is of significance in maintaining blood pressure after dehydration. Aisenbrey et al. (1981) noted that intravenous administration of dPTyr(Me)AVP to rats after 24 h fluid deprivation lowered blood pressure by 14 mmHg. The fall in blood pressure was primarily due to a fall in peripheral resistance. Mean plasma vasopressin was 21.6 pg/ml as measured in a separate group of water-deprived rats. Similar results were obtained by Andrews and Brenner (1981) using a different vasopressin antagonist in anesthetized rats. These authors reported that with the renin-angiotensin system intact, $d(CH_2)_5$ VDAVP caused a transient fall in blood pressure in water-deprived rats. Inhibition of angiotensin II-induced vasoconstriction, by infusion of the competitive antagonist saralasin or removal of the main source of renin following nephrectomy, prevented the return of blood pressure to normal which was otherwise observed following administration of $d(CH_2)_5$ VDAVP. These findings point to an important role of vasopressin as a pressor hormone during fluid deprivation. However, other authors could not confirm these findings (Fejes-Toth et al. 1982; Rascher et al., manuscript in preparation). The shortcomings in the study of Andrew and Brenner (1981) are the use of anesthetized animals and the lack of vasopressin measurements. Under anesthesia, vasopressin is elevated and may cause pressor effects (Bonjour and Malvin 1970). Unfortunately, Aisenbrey et al. (1981) measured blood pressure 1 after completing surgery for catheter implantation which was performed with ether anesthesia. Trying to reproduce their experiments, we found that dPTyr(Me)AVP did indeed lower arterial blood pressure in water-deprived rats by 6 mm Hg 1 h after surgery, but did not affect blood pressure 4 h after surgery (Fig. 9). Plasma concentration of vasopressin 1 h after surgery in water-deprived rats was 44 ± 6.4 pg/ml, whereas it was 12.3 ± 2.2 pg/ml 4 h after surgery (Fig. 10). These results indicate that studies with vasopressin antagonists should be performed in rats conscious after full recovery from anesthesia. In addition, plasma vasopressin concentrations should be measured in the same animals or in a separate group of animals under identical conditions. It is still unclear whether vaso-

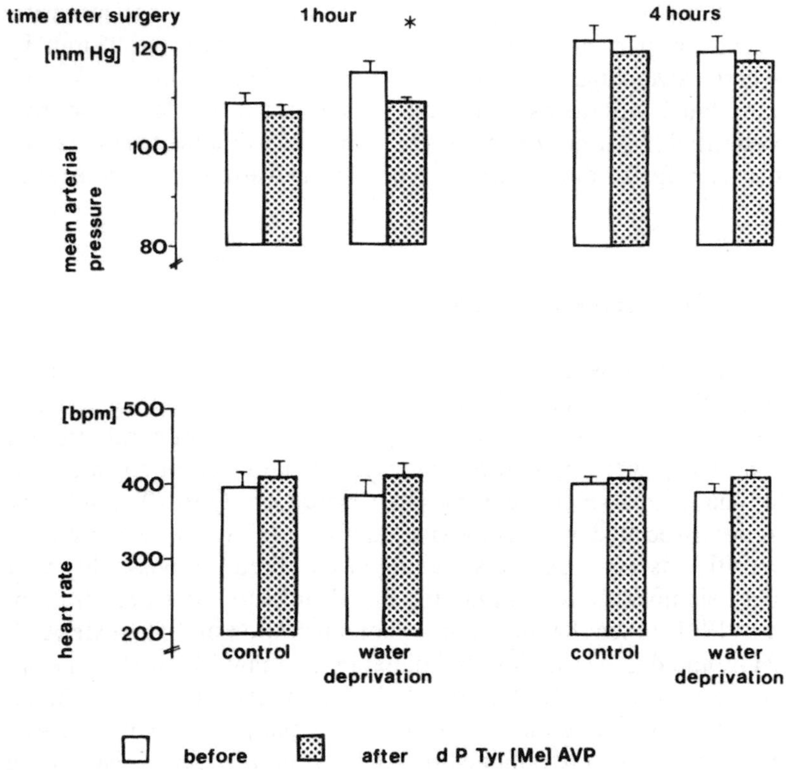

Fig. 9. Changes in mean arterial blood pressure and heart rate following intravenous injection of the vascular vasopressin antagonist dPTyr(Me)AVP (30 μg/kg) in water-deprived (24 h) and control rats 1 and 4 h after completion of surgery for catheter implantation. Note that the antagonist only lowered blood pressure significantly 1 h after completion of surgery in water-deprived rats. *, $p < 0.01$

pressin has a role in blood pressure control after dehydration. Further studies should include measurements of cardiac output or changes in regional blood flow.

In contrast to dehydration, hemorrhage increases vasopressin levels in the circulation far above those required to produce maximal antidiuresis (Weinstein et al. 1960; Share 1961; Rocha e Silva and Rosenberg 1969; Claybough and Share 1973; Arnauld et al. 1977; Larsson et al. 1978; Cowley et al. 1980; Weitzman et al. 1980).

It has been demonstrated in dogs that when the neurohypophysis is removed, recovery from hemorrhage is hampered (Frieden and Keller 1954). Experiments reported by Cowley et al. (1980) provided definite evidence that vasopressin plays a role maintaining blood pressure after hemorrhage in dogs. After bilateral nephrectomy and removal of efferent sympathetic nervous acitivity by spinal cord transsection, the compensatory rise in arterial blood pressure after hemorrhage, which occurred within 3 min, could be abolished by removing the pituitary gland

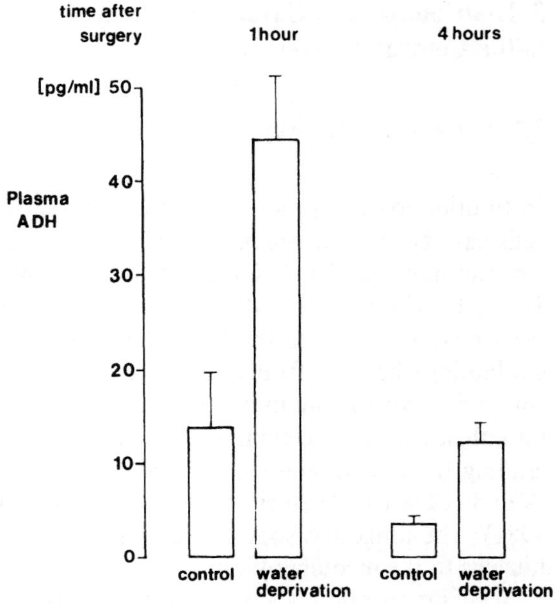

Fig. 10. Plasma vasopressin (ADH) concentration in water-deprived (24 h) and control rats 1 and 4 h hafter completion of surgery for catheter implantation

or by intravenous injection of the vascular vasopressin antagonist dPVDAVP. These authors concluded that vasopressin has potent systemic vasoconstrictor actions which contribute significantly to the restoration of arterial pressure during hemorrhage. Laycock et al. (1979) showed that Brattleboro rats, homozygous for a hypothalamic defect in vasopressin synthesis, were less efficient in compensating arterial blood pressure after hemorrhage than normal Long-Evans rats.

Studying the sympathetic nervous system and the renin-angiotensin system during hemorrhage in Brattleboro rats, Zerbe et al. (1982) suggested that the subnormal blood pressure recovery in these rats was directly related to vasopressin deficiency and was not due to secondary abnormalities of other pressor system.

In conscious dogs mild hemorrhage which was not hypotensive in intact dogs resulted in a marked decrease in mean arterial blood pressure of 32 mmHg after vascular vasopressin receptors were blocked by $d(CH_2)_5$ AVP and $d(CH_2)_5$ Tyr-(Me)AVP (Schwartz and Reid 1981). Under these conditions heart rate, which did not change significantly in intact dogs, rose by 59 beats/min.

There is thus no doubt that vasopressin plays an important role in maintaining blood pressure during blood volume reduction induced by hemorrhage. The amounts of circulating vasopressin under these conditions (more than 100 pg/ml) are well known to induce a rise in blood pressure. This can be reproduced if similar levels of vasopressin are reached by infusion of vasopressin in normotensive dogs (Pullan et al. 1980). In rats with glycerol-induced renal failure, plasma vasopressin rose to more than 100 pg/ml. Blood pressure was 20 mmHg higher in glycerol-treated rats than in control rats, and could be acutely lowered by the administration of a vasopressin antiserum (Hofbauer et al. 1977) or by a vascular vasopressin antagonist (Hofbauer 1980).

3 Distribution and Cardiovascular Actions of Vasopressin in the Central Nervous System

3.1 Vasopressin Pathways

In addition to the supraoptic and paraventricular nuclei, vasopressin – containing cells have been described in the suprachiasmatic nucleus (Vandesande et al. 1975; Zimmerman et al. 1977; Kirsch 1978; Leeuwen et al. 1978; Sofroniew and Weindl 1978a, b; Rhodes et al. 1981). As yet no vasopressin-containing neurons have been observed outside the hypothalamus, but extrahypothalamic vasopressin-containing fibers rise from hypothalamic cells. The presence of separate vasopressin- and oxytocin-containing pathways in the brain has been demonstrated by immunohistochemical techniques in various extrahypothalamic brain regions, including brain stem and spinal cord (Swanson 1977; Buijs 1978; Sofroniew and Weindl 1978a, b; Nilaver et al. 1980; Sofroniew and Schrell 1981; Zimmerman 1981). The longest vasopressin-containing axons lead from the paraventricular nucleus to the medulla oblongata and to the spinal cord.

The supraoptic nucleus mainly projects to the posterior pituitary; whereas vasopressin-containing fibers originating from the suprachiasmatic nucleus almost exclusively travel to brain regions such as the septum, thalamus, lateral habenula, and nucleus tractus solitarii (Sofroniew and Weindl 1978a, b; Weindl and Sofroniew 1978). The paraventricular nucleus subserves both pituitary and extrahypothalamic neurons, e.g., the hippocampus, amygdala, dorsal motor nucleus of the vagus, nucleus tractus solitarii, nucleus ambiguus, and substantia gelatinosa of the spinal cord (Buijs 1978; Nilaver et al. 1980). It is interesting to note that there are more oxytocin-containing than vasopressin-containing neurons projecting to the cardiovascular centers in the brain stem. Electron-microscopic studies have revealed that extrahypothalamic vasopressin exists in presynaptic boutons, which points to a potential neuromodulator role of this hormone (Buijs and Swaab 1979). Release of vasopressin and oxytocin in the brain after depolarization by high potassium and veratridine has been shown and suggests that central vasopressin and oxytocin release is a synaptic event (Buijs and von Heerikhuize 1982).,

Early efforts to quantify vasopressin in extrahypothalamic areas of the brain failed, probably because of insensitive assay methods or inadequate extraction procedures (George and Jacobowitz 1975; Rosenbloom and Fisher 1975). Subsequently, the immunohistochemical findings were corroborated by quantitative measurements using sensitive radioimmunoassays (Dogterom et al. 1978; Glick and Brownstein 1980; Hawthorn et al. 1980; Lang et al. 1981; Rossor et al. 1981). Remarkable concentrations of vasopressin and oxytocin were measured in the medulla oblongata. Vasopressin and oxytocin in the rat brain were identified by means of high-pressure liquid chromatography in combination with radioimmunoassay (Fig. 11) (Lang et al. 1982; Rascher et al. 1982c; Lang et al. 1983). Lang et al. (1983) recently reported that lesions of the paraventricular nucleus of the hypothalamus resulted in a 50% decrease of vasopressin in the brain stem, whereas oxytocin in this brain area disappeared almost completely. It was con-

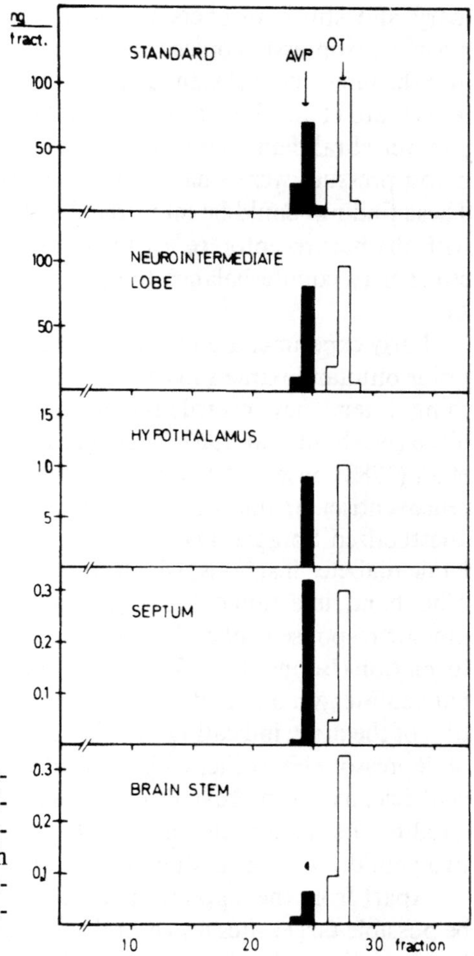

Fig. 11. Characterization of arginine-vasopressin (AVP) and oxytocin (OT) from various brain regions of rats by high-pressure liquid chromatography in combination with radioimmunoassay. Note that immunoreactive material corresponds to synthetic standards (top *panel*). (Lang et al. 1982)

cluded that in contrast to oxytocin, the paraventricular nucleus is not the only site of origin for vasopressin fibers projecting to the brain stem. Little is known about the functional role of extrahypothalamic vasopressin and oxytocin axons. Central vasopressin has been implicated in the release of adrenocorticotropic and growth hormones (Yates et al. 1971; Brodish and Lymangrover 1977; Carlson et al. 1982) and has been related to behavioral effects (De Wied and Versteeg 1979), particularly to memory consolidation (Kovacs et al. 1979). Furthermore, a role of vasopressin as transmitter in antipyresis has been suggested (Veale et al. 1981). Recently, it has been shown that vasopressin exerts a central dipsogenic effect (Szczepanska-Sadowska 1982).

3.2 Cardiovascular Effects of Vasopressin in the Central Nervous System

The intracellular localization of vasopressin and oxytocin in and around important cardiovascular control centers suggests a role of these peptides in the central regulation of arterial blood pressure, although the experimental data are preliminary and still controversial. Evidence for a central cardiovascular action of plasma vasopressin comes from experiments comparing infusions of the peptide into the inferior caval vein of conscious dogs, with infusions into the vertebral artery (Liard et al. 1981). In this study, heart rate decreased to a greater extent after intravertebral than after intravenous drug infusion; however, increases in arterial blood pressure were smaller after intravertebral administration of vasopressin. These findings could be interpreted to mean that vasopressin interferes centrally with the baroreceptor reflex to decrease cardiac output. This central depressor effect may counterbalance the peripheral vasoconstrictor activity of vasopressin.

Early experiments employing intracerebroventricular injections of crude posterior pituitary extracts in cats and dogs did not result in clear-cut effects. If anything, a tendency towards blood pressure decrease and bradycardia was recognized (Nashold et al. 1962; Varma et al. 1969). However, in a recent study Pittman et al. (1982) observed a rise in blood pressure of short latency following intracerebroventricular injections of vasopressin in doses from 25 pmol to 5 μmol in anesthetized Sprague-Dawley rats, as well as in rats homozygous for hypothalamic diabetes insipidus, which completely lack endogenous vasopressin. On the other hand, injection of vasopressin into the lateral brain ventricle attenuated the pressor response evoked by electrical stimulation of the mesencephalic reticular formation (Bohus et al. 1982). Microinjections of vasopressin into the hippocampal dentate gyri imitated the effect seen after intracerebroventricular administration of the drug, indicating that this brain region may play a role in the mediation of depressor effects elicited by central vasopressin. When injected into the fourth ventricle, oxytocin proved to be even more potent, whereas vasopressin administered by the same route had no effect. This points to different target sites of the two peptides for the modulation of cardiovascular regulation.

Apart from the hippocampus, brain stem nuclei have also been speculated to be possible target sites for centrally acting vasopressin. Among these, the nucleus tractus solitarii, the first synapse of the afferent limb of the baroreceptor reflex arch, has attracted particular attention, since it has been shown to receive vasopressin-containing fibers from the suprachiasmatic and paraventricular nuclei (Sofroniew and Weindl 1978b; Buijs 1978; Nilaver et al. 1980). On the basis of indirect evidence, it was postulated that circulating vasopressin enhances reflex bradycardia and the reflex-induced reduction of cardiac output by a central mechanism (Montani et al. 1980; Liard 1980). In contrast, Ciriello and Calaresu (1980) demonstrated that direct stimulation of the paraventricular and supraoptic nuclei in the cat inhibited the reflex vagal bradycardia elicited by stimulation of the carotid sinus nerve. These authors suggested that vasopressin-synthesizing nuclei maintain a tonic inhibitory influence on the carotid sinus reflex.

In agreement with this hypothesis is the finding that microinjections of vasopressin into the nucleus tractus solitarii of anesthetized rats revealed dose-dependent increases in blood pressure (Matsuguchi et al. 1982; Fink et al. 1982); whereas a marked depressor effect was found after microinjection of the vasopressin antagonist $d(CH_2)_5$ AVP (Rascher et al. 1982). Thus, while vasopressin appears to inhibit pressor effects in the hipocampus, it has blood pressure-increasing effects in the brain stem and in periventricular regions.

Vasopressin may thus exert central pressor and depressor effects. These findings may be logically explained by the fact that hormone release can be stimulated by both isosmotic hypovolemia and normo- or hypervolemic hyperosmolality. It is tempting to speculate that isosmotic hypovolemic stimulation (e.g., hemorrhage) results not only in a release of vasopressin into the periphery and a hormonal action on the kidney and blood vessels, as discussed above, but also in an activation of central pressor pathways. Depressor pathways may be activated in hyperosmolar states without volume deficits, such as saline overloading, since under these conditions blood pressure will be normal or already high. This hypothesis can be tested.

3.3 Brain Vasopressin in Experimental Hypertension

The possible role of brain vasopressin in hypertensive mechanisms was investigated in SHR. In these rats, vasopressin content was found to be low in the brain stem, the hypothalamus, and the septum pellucidum, but was not reduced in the pituitary gland as compared to normotensive Wistar-Kyoto rats (Möhring et al. 1980b; Lang et al. 1981, 1982; Morris et al. 1981; Rascher et al. 1982c). The reduction of vasopressin may indicate either reduced synthesis or enhanced release of this peptide. The low plasma levels (Rascher et al. 1981b), together with the diminished hypothalamic concentrations, suggest an impaired synthesis. Such a view is further supported by the finding that after 48 h dehydration, plasma vasopressin rose similarly in both SHRSP and normotensive Wistar-Kyoto rats. However, the pituitary vasopressin content was significantly lower in SHRSP (Rascher et al. 1982b). This indicates reduced synthesis of vasopressin or diminished neurosecretory transport of the hormone from the hypothalamus to the neurohypophysis. Morris and Keller (1982) found a significant reduction of vasopressin in the paraventricular nucleus, but not in the supraoptic nucleus of SHR compared to normotensive Wistar-Kyoto rats. However, Fink, Lang, and Rascher (unpublished observation) could not confirm these findings. They found a similar reduction of vasopressin in both paraventricular and supraoptic nuclei of SHRSP compared to Wistar-Kyoto rats. The mechanism(s) underlying the reduced levels of vasopressin and oxytocin in the various parts of the brain (Lang et al. 1982), or its pathophysiological significance, is not clear. A genetic defect completely unrelated to blood pressure regulation cannot be excluded as a possible cause for the low vasopressin concentrations.

The results on changes in brain tissue content of vasopressin in SHR do not exclude the possibility that central vasopressin acts as a local depressor hormone and that the reduced vasopressin content in brains of SHRSP may contribute to the hypertension of these animals by leaving other pressor mechanisms unopposed. The establishment of a new line of hypertensive rats (SHRDI) by crossing SHRSP and rats homozygous for diabetes insipidus (by Ganten et al. 1983) could be interpreted to support such a hypothesis and allows two conclusions: First, spontaneous hypertension can develop in the absence of vasopressin. Second, in view of depressor effects of brain vasopressin (Bohus et al. 1982) this genetic defect may even facilitate the establishment of hypertension, since other pressor

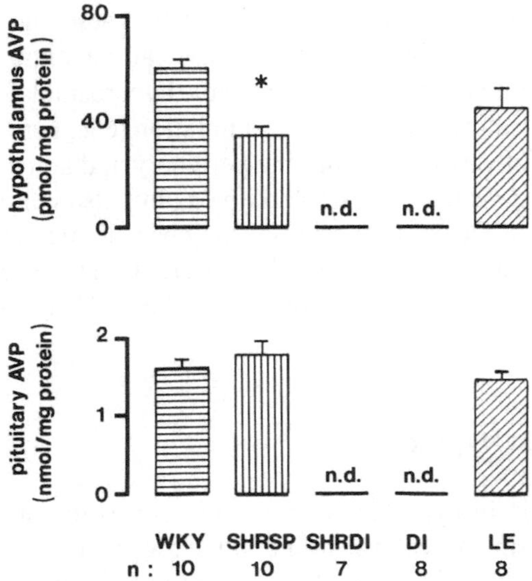

Fig. 12. Content of arginine-vasopressin (AVP) in the hypothalamus and pituitary gland of SHRDI compared to various age-matched control groups. WKY, Wistar-Kyoto rats: DI, diabetes insipidus rats; LE Long-Evans rats. *, $p<0.05$ (Ganten et al. 1983)

stimuli, such as increased sympathetic tone (Schömig et al. 1978), remain unopposed in the vasopressin-deficient SHRDI. The high blood pressure in SHRDI may thus occur not *despite* the fact that vasopressin is absent; but this defect may even *contribute* to the development of high blood pressure.

In SHRDI with complete lack of AVP synthesis (Fig. 12), the level of oxytocin in the pituitary and in the hypothalamus was reduced, as in Brattleboro diabetes insipidus rats (Fig. 13). It has previously been shown that pituitary oxytocin synthesis is intact in diabetes insipidus rats, but that the level of oxytocin in these rats is about one-third that in Long-Evans rats (Valtin et al. 1965). This finding is confirmed by our experiments. In the hypothalamus of diabetes insipidus rats, oxytocin level has only half that found in Long-Evans rats. SHRDI had equally reduced levels in the hypothalamus and pituitary gland compared to normotensive Wistar-Kyoto and Long-Evans rats.

Another possible explanation for the reduced vasopressin content in various brain areas of SHRSP is that these changes are secondary to alterations in the activity of the central sympathetic nervous system. Alterations of central catecholaminergic pathways have been observed in SHR (Versteeg et al. 1976; Saavedra et al. 1978; Fuxe et al. 1979a, b; Wijnen et al. 1980). Fuxe et al. (1979a) have shown that the content of dopamine and norepinephrine is reduced in the hypothalamus of SHRSP. A similar reduction in norepinephrine content was found in the caudal medulla oblongata of SHRSP as compared to normotensive Wistar-Kyoto rats. This finding would be consistent with the hypothesis that alterations in the vasopressin system are secondary to changes in catecholamine pathways.

There is also evidence that vasopressin affects the metabolism of catecholamines, which are themselves involved in blood pressure regulation. Tanaka et al. (1977a, b) observed an enhance norepinephrine and epinephrine turnover in discrete nuclei of rat brain following intracerebroventricular injections of vasopressin. Microiontophoretic administration of vasopressin into the locus ceruleus

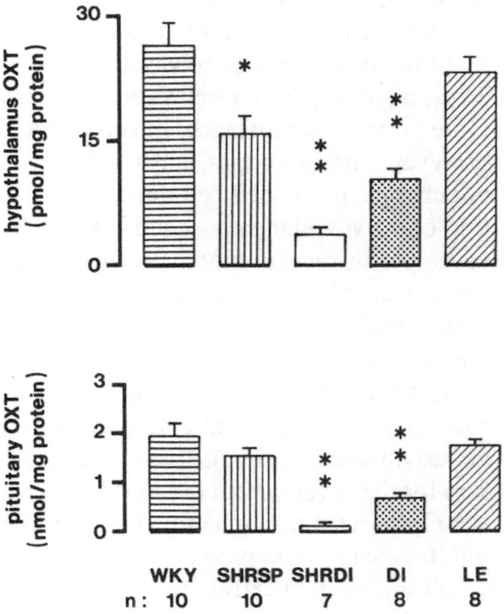

Fig. 13. Content of oxytocin in the hypothalamus and pituitary gland of SHRDI rats compared to various age-matched control rats (same animals as Fig. 12). *, $p<0.05$; **, $p<0.01$

of rats activates noradrenergic neurons in this particular brain area (Olpe and Baltzer 1981). Recently, Blessing et al. (1982) demonstrated in rabbits that hypertension following destruction of noradrenergic neurons in brain stem is mediated by vasopressin release. Furthermore, these results are consistent with the hypothesis that noradrenergic neurons (A_1 cell group) tonically inhibit the activity of vasopressin-secreting neuroendocrine cells.

In summary, the exact role of vasopressin fibers in cardiovascullar control centers of the brain stem is far from being clear. However, evidence has been presented that they may be involved as neurotransmitters or neuromodulators in the central regulation of arterial blood pressure. It is not unusual that the same neurotransmitter system can activate both blood pressure-lowering and blood pressure-increasing pathways. This is well known in catecholamines, centrally and peripherally, and a similar concept emerges for the opioid peptides (Lord et al. 1977; Ganten et al. 1981). As is well known from catecholamines, the pressor and depressor activities of a neurotransmitter system are determined not only by its different transmitter molecules (dopamine, norepinephrine), but also by different receptors and their possible pre- and postsynaptic localization. Different receptors of vasopressin are already known in the periphery (Jard and Bockaert 1975; Penit et al. 1983), and it can be expected that central vasopressin and oxytocin receptors will be described in the near future.

4 Summary

The two major biological actions of vasopressin are antidiuresis and vasoconstriction. The antidiuretic action of low concentrations of vasopressin is well established, whereas concentrations 10–100 times above those required for antidiuresis elevate arterial blood pressure. Recent experiments have provided evidence that low concentrations of vasopressin do have significant hemodynamic effects. Thus, physiological concentrations of vasopressin cause vasoconstriction and elevate peripheral resistance. In subjects with intact cardiovascular reflex activity, however, cardiac output falls concomitantly and blood pressure therefore does not change. It is concluded that vasopressin is a potent vasoconstrictor (contraction of the vascular smooth muscle) and is therefore capable of determining local tissue perfusion and blood flow distribution. In the intact organism, the vasopressor activity and blood pressure-increasing potency of this hormone are largely buffered by cardiovascular reflex regulation. This indicates that the cardiovascular actions of vasopressin are more complex than previously assumed.

In DOCA-hypertensive rats, vasopressin appears to have at least a permissive role for the development and maintenance of hypertension, whereas in spontaneously hypertensive rats there is definite evidence that vasopressin is not necessary for the development of high blood pressure. During hemorrhage, vasopressin is an important hormone for the restoration of blood pressure to normal values, but its role in blood pressure control during dehydration is not clear.

Vasopressin in extrahypothalamic areas of the brain affects circulatory regulation by interactions with central cardiovascular control centers. The exact mechanism of how brain vasopressin is involved in the central regulation of blood pressure remains to be established, but is presently a field of active research.

References

Aisenbrey GA, Berl T (1982) Role of vasopressin in the control of systemic hemodynamics – lessons learned from the Brattleboro rat. Ann NY Acad Sci 394:299–307

Aisenbrey GA, Handleman WA, Arnold P, Manning M, Schrier RW (1981) Vascular effects of arginine vasopressin during fluid deprivation in the rat. J Clin Invest 67:961–968

Altura BM (1973) Selective microvascular constrictor actions of some neurohypophyseal peptides. Eur J Pharmacol 24:49–60

Altura BM (1975) Dose-response relationships for arginine vasopressin and synthetic analogues on three types of rat blood vessels: possible evidence for regional differences in vasopressin receptor sites within a mammal. J Pharmacol Exp Ther 193:413–423

Altura BM, Altura BT (1977) Vascular smooth muscle and neurohypophyseal hormones. Fed Proc 36:1853–1860

Andrews CE, Brenner BM (1981) Relative contribution of arginine vasopressin and angiotensin II to maintenance of systemic arterial pressure in the anesthetized water-deprived rat. Circ Res 48:254–258

Arnauld F, Czernichow P, Fumoux F, Vincent JD (1977) The effects of hypotension and hypovolemia on the liberation of vasopressin during hemorrhage in the unanesthetized monkey. Pfluegers Arch 371:193–200

Bankowski K, Manning M, Haldar J, Sawyer WH (1978) Design of potent antagonists of the vasopressor response to arginine-vasopressin. J Med Chem 21:850–853

Bartelstone HJ, Nasmyth PA (1965) Vasopressin potentiation of catecholamine actions in dog, rat, cat and rat aortic strip. Am J Physiol 208:754–762

Berde B, Boissonnas RA (1968) Basic pharmacologic properties of synthetic analogues and homologues of the neurohypophyseal hormones. In: Berde B (ed) Neurohypophysial hormones and similar polypeptides, Springer, Berlin Heidelberg New York (Handbook of Experimental Pharmacology 23, pp 802–870)

Berde B, Hugnenin RL, Stürmer E (1964) Vasopressin analogues with selective pressor activity. Experimentia 20:42–43

Berecek KH, Murray RD, Gross F, Brody MJ (1982) Vasopressin and vascular reactivity in the development of DOCA hypertension in rats with hereditary diabetes insipidus. Hypertension 4:3–12

Blessing WW, Sved AF, Reis DJ (1982) Destruction of noradrenergic neurons in rabbit brainstem elevates vasopressin, causing hypertension. Science 217:660–661

Bohus B, Versteeg CAM, De Jong W (1982) Vasopressin and central blood pressure control in the rat. In: Rascher W, Clough D, Ganten D (eds), Hypertensive mechanisms. Schatthauer, Stuttgart, pp 592–596

Bonjour JP, Malvin RL (1979) Plasma concentration of ADH in conscious and anesthetized dogs. Am J Physiol 218:1128–1132

Brazeau P (1975) Agents affecting the renal conservation of water. In: Goodman LS, Gilman A (eds) The pharmacological basis of therapeutics 5; MacMillan, New York, pp 849–859

Bridges TE, Thorn NA (1970) The effect of autonomic blocking agents on vasopressin release in vivo induced by osmoreceptor stimulation: J Endocrinol 48:265–276

Brodish A, Lymangrover JR (1977) The pituitary adrenocortical system. In: McCann MD (eds) International review of physiology, endocrine physiology II, vol 16. University Park Press, Baltimore, pp 93–149

Brownstein MJ, Russell JT, Gainer H (1980) Synthesis, transport and release of posterior pituitary hormones. Science 207:373–378

Buijs RM (1978) Intra- and extrahypothalamic vasopressin and oxytocin pathways in the rat. Cell Tissue Res 192:423–435

Buijs RM, Swaab DF (1979) Immuno-electron microscopical demonstration of vasopressin and oxytocin synapses in the limbic system of the rat. Cell Tissue Res 204:355–365

Buijs RM, van Heerikhuize JJ (1982) Vasopressin and oxytocin release in the brain – a synaptic event. Brain Res 252:71–76

Carlson DE, Dornhorst A, Seif SM, Robinson AG, Gann DS (1982) Vasopressin-dependent and vasopressin-independent control of the release of adrenocorticotropin. Endocrinology 110:680–682

Chenoweth MB, Ellman GL, Reynolds RC, Shea PJ (1958) A secondary response to pressor stimuli caused by sensitization to an endogenous pituitary hormone. Circ Res 6:334–342

Ciriello J, Calaresu FR (1980) Role of paraventricular and supraoptic nuclei in central cardiovascular regulation in the cat. Am J Physiol 239:R123–R140

Claybough JR, Share L (1973) Vasopressin, renin and cardiovascular responses to continuous slow hemorrhage. J Physiol (Lond) 224:519–523

Cowley AM Jr, Monos E, Guyton AC (1974) Interaction of vasopressin and the baroreceptor reflex system in the regulation of arterial blood pressure in the dog. Circ Res 34:505–514

Cowley AM, Switzer SJ, Guinn MM (1980) Evidence and quantification of the vasopressin arterial pressure control system in the dog. Circ Res. 46:58–67

Cowley AW, Cushman WC, Quillen EW, Skelton MM, Langford HG (1981) Vasopressin elevation in essential hypertension and increased responsiveness to sodium intake. Hypertension 3 [Suppl I]:I-93-I-100

Crofton JT, Share L, Shade RE, Allen C, Tarnowski D (1978) Vasopressin in the rat with spontaneous hypertension. Am J Physiol 235:H361–H366

Crofton JT, Share L, Shade RE, Lee-Kwon WJ, Manning M, Sawyer WH (1979) The importance of vasopressin in the development and maintenance of DOC-salt hypertension in the rat. Hypertension 1:31–38

Crofton JT, Share L, Wang BC, Shade RE (1980) Pressor responsiveness to vasopressin in the rat with DOC-salt hypertension. Hypertension 2:424–431

Crofton JT, Share L, Baer PG, Allen CM, Wang BC (1981) Vasopressin secretion in the New Zealand genetically hypertensive rat. Clin Exp Hyperten 3:975–989

De Wied D, Versteeg DHG (1979) Neurohypophyseal principles and memory. Fed Proc 38:2348–2354

Di Pette DJ, Gavras I, North WG, Brunner HR, Gavras H (1982) Vasopressin in salt-induced hypertension of experimental renal insufficiency. Hypertension 4 [Suppl II]:II-125-II-130
Dogterom J, Snijdewint FGM, Buijs RM (1978) The distribution of vasopressin and oxytocin in the rat brain. Neurosci Lett 9:341–346
Dousa TP, Valtin H (1976) Cellular actions of vasopressin in the mammalian kidney. Kidney Int 10:46–63
Dreifuss JJ (1975) A review of neurosecretory granules: their contents and mechanism of release. Ann NY Acad Sci 248–184–201
Dunn FJ, Brennan TJ, Nelson AE, Robertson GL (1973) The role of blood osmolality and volume in regulating vasopressin secretion in the rat. J Clin Invest 52:3212–3219
Du Vigneuad V, Gish DI, Katsoyannis PG (1954a) A synthetic preparation possessing biological properties associated with arginine-vasopressin. J Am Chem Soc 76:4751–4752
Du Vigneaud V, Ressler C, Swan JM, Katsoyannis PG, Roberts CW (1954b) The synthesis of oxytocin. J Am Chem Soc 76:3115–3121
Ellis ME, Grollman A (1949) The antidiuretic hormone in the urine in experimental and clinical hypertension. Endocrinology 44:415–419
Erker EF, Chan WY (1977) The site and mechanism of phenoxybenzamine potentiation of the pressor response to oxytocin and vasopressin: In vivo and isolated aortic strips studies. J Pharmacol Exp Ther 202:287–293
Farini F (1913) Diabete insipido ed opoterapia. Gazz Osped Clin 34:1135–1139
Fejes-Toth G, Naray-Fejes-Toth A, Ratge D, Frölich JC (1982) Vasopressin released by osmotic stimuli does not contribute to blood pressure regulation in the rat. Therapiewoche 32:5723
Fink B, Rascher W, Rettig R, Gross F (1982) Anstieg von Blutdruck und Herzfrequenz nach Mikroinjektion von Vasopressin in den Nucleus tractus solitarii. Therapiewoche 32:5724
Frieden J, Keller AD (1954) Decreased resistance to hemorrhage in neuro-hypophysectomized dogs. Circ Res 2:214–220
Friedman SM, Friedman CL, Nakashima M (1960) Accelerated appearance of DCA hypertension in rats treated with pitressin. Endocrinology 67:752–759
Fuxe K, Ganten D, Jonsson G, Agnati LF, Andersson K, Hökfelt T, Bolme P, Goldstein M, Hallman H, Unger Th, Rascher W (1979a) Catecholamine turnover changes in hypothalamus and dorsal midline area of the caudal modedulla oblongata of spontaneously hypertensive rats. Neurosci Lett 15:283–288
Fuxe K, Ganten D, Jonsson G, Bolme P, Agnati LF, Andersson K, Goldstein M, Hökfelt T (1979b) Evidence for a selective reduction of adrenaline turnover in the dorsal midline area of the caudal medulla oblongata of young spontaneously hypertensive rats. Acta Physiol Scand 107:397–399
Gainer H, Sarne Y, Brownstein MJ (1977) Neurophysin biosynthesis: Conversion of a putative precursor during axonal transport. Science 195:1354–1356
Ganten D, Unger T, Schölkens B, Rascher W, Speck G, Stock G (1981) Role of neuropeptides in regulation of blood pressure. In: Abboud FM, Fozzard HA, Gilmore JC, Reis DJ (eds) Disturbances in neurogenic control of the circulation. Am Physiol Soc, Washington, pp 139–151
Ganten U, Rascher W, Lang RE, Dietz R, Rettig G, Unger T, Taugner R, Ganten D (1983) Development of a new strain of spontaneously hypertensive rats homozygous for hypothalamic diabetes insipidus. Hypertension 5 [Suppl I]:I-119-I-128
Gauer OH, Henry JP, Behn C (1970) The regulation of extracellular fluid volume. Annu Rev Physiol 32:547–595
Gavras H, Hatzinikolaou P, North WG, Bresnahan M, Gavras I (1982) Interaction of the sympathetic nervous system with vasopressin and renin in the maintenance of blood pressure regulation. Hypertension 4:400–405
George JM, Jacobowitz DM (1975) Localization of vasopressin in discrete areas of the rat hypothalamus. Brain Res 93:363–366
Glick SM, Brownstein MJ (1980) Vasopressin content of rat brain. Life Sci 27:1103–1110

Gluck S, Al-Awqati Q (1980) Vasopressin increases water permeability by inducing pores. Nature 284:631–632

Goetz TL, Bond GL, Bloxham DD (1975) Atrial receptors and renal function. Physiol Rev 55:157–205

Harmanci MC, Stern P, Kackadorian WA, Valtin H, Di Scala VA (1980) Vasopressin and collecting duct intamembranous particle clusters: a dose-response relationship. Am J Physiol 239:F560–F564

Hatzinikolaou P, Gavras H, Brunner HR, Gavras I (1980) Sodium-induced elevation of blood pressure in the anephritic state. Science 209:935-936

Hatzinikolaou P, Gavras H, Brunner HR, Gavras I (1981) Role of vasopressin, catecholamines and plasma volume in hypertonic saline-induced hypertension. Am J Physiol 240:H827–H831

Hawthorn K, Ang VTY, Jenkins JS (1980) Localization of vasopressin in the rat brain. Brain Res 197:75–81

Hayward JN, Pavasuthipaisit K, Perez-Lopez FR, Sofroniew MV (1976) Radioimmunoassay of arginine-vasopressin in rhesus monkey plasma. Endocrinology 98:975–981

Heyndrickx G, Boettcher D, Vatner S (1976) Effects of angiotensin, vasopressin and methoxamine on cardiac function and blood flow distribution in conscious dogs. Am J Physiol 231:1579–1587

Hofbauer KG (1980) Effect of a competitive antagonist of arginine-vasopressin in glycerol-induced acute renal failure in rats. Naunyn Schmiedebergs Arch Pharmacol 311 [Suppl):R50

Hofbauer KG, Konrads A, Bauereiss K, Möhring B, Möhring J, Gross F (1977) Vasopressin and renin in glycerol-induced acute renal failure in the rat. Circ Res 41:424–428

Hoffman WE, Phillips MI, Schmid PG, Falcon J, Weet JF (1977a) Antidiuretic hormone release and the pressor response to central angiotensin II and cholinergic stimulation. Neuropharmacology 16:463–472

Hoffman EW, Phillips MI, Schmid P (1977b) The role of catecholamines in central antidiuretic and pressor mechanism. Neuropharmacology 16:563–569

Howell WH (1898) The physiological effects of extracts of the hypophysis cerebri and infundibular body. J Exp Med 3:245–258

Ishikawa S, Schrier RW (1982) Evidence for a role of opioid peptides in the release of arginine vasopressin in the conscious rat. J Clin Invest 69:666–672

Ishikawa S, Handleman WA, Schrier RW, Berl T (1982) In vivo effect of cellular calcium uptake on osmotic and nonosmotic release of arginine vasopressin. Kidney Int 21:813–819

Iwamoto HS, Rudolph AM, Keil LC, Heyman MA (1979) Hemodynamic responses to the sheep fetus to vasopressin infusion. Circ Res 44:430–436

Jard S, Bockaert J (1975) Stimulus response coupling in neurohypophysical peptide target cells. Physiol Rev 55:489–536

Johnson MD, Kinter LB, Beeuwkes R (1979) Effect of AVP and DDAVP on plasma renin activity and electrolyte excretion in conscious dogs. Am J Physiol 236:F66–F70

Johnson JA, Ichikawa S, Kurz KD, Fowler WL, Payne CG (1981) Pressor responses to vasopressin in rabbits with 3-day renal artery stenosis. Am J Physiol 240:H862–H867

Johnston CI, Newman M, Woods R (1981) Role of vasopressin in cardiovascular homeostasis and hypertension. Clin Sci 61:129s–139s

Kamm O, Aldrich TB, Corote FW, Row LW, Bugbee EP (1928) Active principles of the posterior lobe of the pituitary gland I. The demonstration of the presence of two active principles II. The separation of the two principles and their concentration in the form of potent solid preparations. J Am Chem Soc 50:573–580

Karashima T (1981) Effects of vasopressin on smooth muscle of guinea pig mesenteric vessels. Br J Pharmacol 72:673–684

Keil LC, Summy-Long J, Severs WB (1975) Release of vasopressin by angiotensin II. Endocrinology 96:1063–1065

Kepinow D (1912) Über den Synergismus von Hypophysisextrakt und Adrenalin. Naunyn Schmiedebergs Arch Pharmacol 67:247–274

Khokhar AM, Slater JDH (1976) Increased renal excretion of arginine-vasopressin (AVP) during mild hydropenia in young men with mild essential benign hypertension. Clin Sci Med 51:691s–694s

Khokhar AM, Slater JDH, Ma J, Ramage CM (1980) The cardiovascular effect of vasopressin in relation to its plasma concentration in man and its relevance to high blood pressure. Clin Endocrinol (Oxf) 13:259–266

Kimura T, Share L, Wang BC, Crofton JT (1981 a) Central effects of dopamine and bromocriptine on vasopressin release and blood pressure. Neuroendocrinology 33:347–351

Kimura T, Share L, Wang BC, Crofton JT (1981 b) The role of central adrenoceptors in the control of vasopressin release and blood pressure. Endocrinology 108:1829–1836

Kirchheim HR (1976) Systemic arterial baroreceptor reflexes. Physiol Rev 56:100–177

Kirsch B (1978) Light- and electron microscopic localisation of vasopressin or vasopressin-like substance in the neurons of the rat suprachiasmatic nucleus. Cell Tissue Res 194:361–365

Kovacs GL, Bohus B, Versteeg DHG (1979) The effects of vasopressin in memory processes: the role of noradrenergic transmission. Neuroscience 4:1529–1537

Kruszynski M, Lammek B, Manning M, Seto J, Haldar J, Sawyer WH (1980) ⟨1-(β-mercapto-β,β-cyclopentamethylenepropionic acid),2-(0-methyl)tyrosine⟩ arginine vasopressin and ⟨1-(β-mercapto-β,β-cyclopentamethylenepropionic acid⟩ arginine vasopressin, two highly potent antagonists of the vasopressor response to arginine vasopressin. J Med Chem 23:364–368

Land H, Schütz G, Schmale H, Richter D (1982) Nucleotide sequence of cloned cDNA encoding bovine arginine vasopressin-neurophysin II precursor. Nature 295:299–303

Lang H, Grez M, Ruppert S, Schmale H, Rehbein M, Richter D, Schütz G (1983) Deduced amino acid sequence from the bovine oxytocin-neurophysin I precursor cDNA. Nature 302:342–344

Lang RE, Rascher W, Unger T, Ganten D (1981) Reduced content of vasopressin in the brain of spontaneously hypertensive as compared to normotensive rats. Neursci Lett 23:199–202

Lang RE, Rascher W, Hermann K, Unger T, Ganten D (1982) Distribution of oxytocin and vasopressin in rat brain: reduced content in spontaneously hypertensive rats. In: Rascher W, Clough D, Ganten D (eds) Hypertensive mechanisms. Schattauer, Stuttgart, pp 581–587

Lang RE, Heil J, Ganten D, Hermann K, Rascher W, Unger T (1983) Effects of lesions in the paraventricular nucleus of the hypothalamus on vasopressin and oxytocin contents in brainstem and spinal cord of rat. Brain Res 260:326–329

Larsson B, Olsson K, Fyhrquist F (1978) Vasopressin release induced by hemorrhage in the goat. Acta Physiol Scand 104:309–317

Laycook JF, Penn W, Shirley DG, Walter SJ (1979) The role of vasopressin in blood pressure regulation immediately following acute hemorrhage in the rat. J Physiol 296:267–275

Lee-Kwon WJ, Share L, Crofton JT, Shade RE (1981) Vasopressin in the rat with partial nephrectomy-salt hypertension. Clin Exp Hypertens 3:281–297

Leeuwen FW, van Swaab DG, De Raaby C (1978) Immunoelectron-microscopic localization of vasopressin in the rat suprachiasmatic nucleus. Cell Tissue Res 193:1–10

Liard JF (1980) The baroreceptor reflexes in experimental hypertension. Clin Exp Hypertens 2:479–498

Liard JF (1982) Vasopressin in cardiovascular control and hypertension. In: Rascher W, Clough D, Ganten D (eds) Hypertensive mechanisms. Schattauer, Stuttgart, pp. 561–573

Liard JF, Deriaz O, Tschopp M, Schoun J (1981) Cardiovascular effects of vasopressin infused into the vertebral circulation of conscious dogs. Clin Sci 61:345–347

Lohmeier TE, Smith MK, Cowley AM, Manning TD, Guyton AC (1981) Is vasopressin an important hypertensive hormone? Hypertension 3:416–425

Lord JAH, Waterfield AA, Hughes J, Kosterlitz HW (1977) Endogenous opioid peptides: multiple agonists and receptors. Nature 267:495–499

Lowbridge J, Manning M, Haldar J, Sawyer WH (1978) ⟨1-(β-mercapto-β, β-cyclopentamethylenepropionic acid), 4-valine,-8-D-arginine⟩ vasopressin, a potent and selective inhibitor of the vasopressor response to arginine vasopressin. J Med Chem 21:313–315

Manning M, Sawyer WH (1982) Antagonists of vasopressin and antidiuretic responses to arginine vasopressin. Ann Intern Med 96:520–522

Manning M, Balaspiri L, Acosta M, Sawyer WH (1973) Solid-phase synthesis of (1-deamino, 4-valine, 8-D-arginine) vasopressin (dV-DAVP), a highly potent and specific antidiuretic agent possessing protracted effects. J Med Chem 16:975–978

Manning M, Lowbridge J, Haldar J, Sawyer WH (1977a) Design of neurohypophyseal peptides that exhibit selective agonistic and antagonistic properties. Fed Proc 36:848–853

Manning M, Lowbridge J, Stier CT, Haldar J, Sawyer WH (1977b) (1-Deaminopenicillamine,-4-valine)-8-D-arginine-vasopressin, a highly potent inhibitor of the vasopressor response to arginine-vasopressin. J Med Chem 20:1228–1230

Manning RD, Guyton AC, Coleman TG, McCaa RE (1979) Hypertension in dogs during antidiuretic hormone and hypertonic saline infusion. Am J Physiol 236:H314–H322

Manning M, Gzronka Z, Sawyer WH (1981) Synthesis of posterior pituitary hormones and hormone analogues. In: Bearwell C, Robinson G (eds) The pituitary. Butterworths, Kent, pp 265–296

Manning M, Klis WA, Olma A, Seto J, Sawyer WH (1982a) Design of more potent and selective antagonists of the antidiuretic responses to arginine-vasopressin devoid of antidiuretic agonism. J Med Chem 25:414–419

Manning M, Ilma A, Klis WA, Kolodziejczykm AM, Seto J, Sawyer WH (1982b) Design of more potent antagonists of the antidiuretic responses to arginine vasopressin. J Med Chem 25:45–50

Matsuguchi H, Schmid PG (1982) Pressor responses to vasopressin and impaired baroreflex function in DOC-salt hypertension. Am J Physiol 242:H44–H49

Matsuguchi H, Schmid PG, van Orden D, Mark AL (1981) Does vasopressin contribute to salt-induced hypertension in the Dahl strain? Hypertension 3:174–181

Matsuguchi H, Sharabi FM, Gordon FJ, Johnson AK, Schmid PG (1982) Blood pressure and heart rate responses to microinjection of vasopressin into the nucleus tractus solitarii region of the rat. Neuropharmacology 21:687–693

McNeill JR (1974) Intestinal vasoconstriction following diuretic-induced volume depletion: role of angiotensin and vasopressin. Can J Physiol Pharmacol 52:829–839

McNeill JR, Stark RD, Greenway CV (1970) Intestinal vasoconstriction after hemorrhage: roles of vasopressin and angiotensin. Am J Physiol 219:1342–1347

Mc Neill JR, Wicox WC, Pang CCY (1976) Vasopressin and angiotensin: reciprocal mechanisms controlling mesenteric conductance. Am J Physiol 232:262–266

Miller TR, Handlman WA, Arnold PE, Mc Donald KM, Molinoff PB, Schrier RW (1979) Effect of central catecholaminergic depletion on the osmotic and nonosmotic stimulation of vasopressin (antidiuretic hormone) in the rat. J Clin Invest 64:1599–1607

Möhring J (1978) Neurohypophyseal vasopressor principle: vasopressor hormone as well as antidiuretic hormone? Klin Wochenschr 56 [Suppl I]:71–79

Möhring J, Möhring B, Petri M, Haack D (1976) Is vasopressin involved in the pathogenesis of malignant desoxycorticosterone hypertension in rats? Lancet 1:170–173

Möhring J, Möhring B, Petri M, Haack D (1977) Vasopressor role of ADH in the pathogenesis of malignant DOC hypertension. Am J Physiol 232:F260–F269

Möhring J, Möhring B, Petri M, Haack D (1978a) Plasma vasopressin concentrations and effects of vasopressin antiserum on blood pressure in rats with malignant two-kidney Goldblatt hypertension. Circ Res 42:17–22

Möhring J, Kintz J, Schoun J (1978b) Role of vasopressin in blood pressure control of spontaneously hypertensive rats. Clin Sci Mol Med 55:247s–250s

Möhring J, Kintz J, Schoun J (1979) Studies on the role of vasopressin in blood pressure control of spontaneously hypertension rats with established hypertension. J Cardiovasc Pharmacol 1:593–608

Möhring J, Glänzer K, Maciel Jr JA; Düsing R, Kramer HJ, Arbogast R, Koch-Weser J (1980a) Greatly enhanced pressor response to antidiuretic hormone in patients with

impaired cardiovascular reflexes due to idiopathic orthostatic hypotension. J Cardiovasc Pharmacol 2:367–376
Möhring J, Schoun J, Kintz J, McNeill R (1980b) Decreased vasopressin content in brain stem of rats with spontaneous hypertension. Naunyn Schmiedebergs Arch Pharmacol 315:83–84
Möhring J, Kintz J, Schoun J, McNeill JR (1981) Pressor responsiveness and cardiovascular reflex activity in spontaneously hypertensive and normotensive rats during vasopressin infusion. J Cardiovasc Pharmacol 3:948–957
Monos E, Cox RH, Robertson LH (1978) Direct effect of physiological doses of arginine vasopressin on the arterial wall in vivo. Am J Physiol 234:H167–H172
Montani JP, Liard JF, Schoun J, Möhring J (1980) Hemodynamic effects of exogenous and endogenous vasopressin at low plasma concentration in conscious dogs. Circ Res 47:346–355
Morris M, Keller M (1982) A specific deficiency in paraventricular vasopressin and oxytocin in the spontaneously hypertensive rat. Brain Res 7:173–176
Morris M, Wren JA, Sundberg DK (1981) Central neural peptides and catecholamines in spontaneously and DOCA-salt hypertension. Peptides 2:207–211
Morton JJ, Garcia del Rio C, Hughes MJ (1982) Effect of acute vasopressin infusion on blood pressure plasma angiotensin II in normotensive and DOCA-salt hypertensive rats. Clin Sci 62:143–149
Nakano J (1974) Cardiovascular responses to neurohypophyseal hormones. In: Greep RO, Astwood EB (eds) The pituitary gland and its neuroendocrine control. Am Physiol Soc, Washington, DC (Handbook of Physiology, sec 7, Endocrinology, vol. IV I, pp 395–442)
Nashold BSA, Mannarino E, Wunderlich M (1962) Pressor-depressor blood pressure responses in the cat after intraventricular injection of drugs. Nature 193:1297–1298
Nestor JJ, Ferger MF, Du Vigneaud V (1975) (1-β-mercapto-β,β-pentamethylenepropionic acid) oxytocin, a potent inhibitor of oxytocin. J Med Chem 18:284–287
Nilaver G, Zimmerman J, Wilkins J, Michaels J, Hoffman D, Silverman AJ (1980) Magnocellular hypothalamic projections to the lower brain stem and spinal cord of the rat. Neuroendocrinology 30:150–158
Oliver G, Schäfer EA (1895) On the physiological actions of extracts of the pituitary body and certain other glandular organs. J Physiol (Lond) 18:277–279
Olpe HR, Baltzer V (1981) Vasopressin activates noradrenergic neurons in the locus coeruleus: a microiontophoretic investigation. Eur J Pharmacol 73:377–378
Padfield PL, Brown JJ, Lever AF, Morton JJ, Robertson JIS (1976) Changes of vasopressin in hypertension: cause or effect? Lancet 1:1255–1257
Padfield PL, Brown JJ, Lever AF, Morton JJ, Robertson JIS (1981) Blood pressure in acute and chronic vasopressin excess. Studies of malignant hypertension and the syndrome of inappropriate antidiuretic hormone secretion. N Engl J Med 304:1067–1070
Page IH, Sweet JE (1937) The effect of hypophysectomy on arterial blood pressure of dogs with experimental hypertension. Am J Physiol 120:238–245
Pang CCY, Wilcox WC, McNeill JR (1979) Hypophysectomy and saralasin on mesenteric vasoconstrictor response to vasopressin. Am J Physiol 236:H200–H205
Penit J, Faure M, Jard S (1983) Vasopressin and angiotensin II receptors in rat aortic smooth muscle in culture. Am J Physiol 244:E72–E82
Pittman QJ, Lawrence D, Mc Lean L (1982) Central effects of arginine vasopressin on blood pressure in rats. Endocrinology 110:1058–1060
Pohjavuori M, Fyhrquist F (1980) Hemodynamic significance of vasopressin in the newborn infant. J Pediatr 97:462–465
Pullan PT, Johnston CI, Anderson WP, Korner PI (1978) The role of vasopressin in blood pressure control and in experimental hypertension. Clin Sci 55:251s–254s
Pullan PT, Johnston CI, Anderson WP, Korner PI (1980) Plasma vasopressin in blood pressure homeostasis and in experimental renal hypertension. Am J Physiol 239:H81–H87
Quillen EW, Cowley AM (1983) Influence of volume changes on osmolality-vasopressin relationship in conscious dogs. Am J Physiol 244:H73–H79

Rabito SF, Carretero OA, Scicli AG (1981) Evidence against a role of vasopressin in the maintenance of high blood pressure in mineralocorticoid and renovascular hypertension. Hypertension 3:34–38

Rascher W, Lang RE, Taubitz M, Meffle H, Unger T, Ganten D, Gross F (1981a) Vasopressin induced increase in total peripheral resistance in deoxycorticosterone acetate hypertension is buffered by the baroreceptor reflex. Clin Sci 61:153s–156s

Rascher W, Weidemann E, Gross F (1981b) Vasopressin in the plasma of stroke-prone spontaneously hypertensive rats. Clin Sci 61:295–298

Rascher W, Fink B, Gross F (1982a) Increased pressor responses following microinjections of arginine vasopressin into the nucleus tractus solitarii of spontaneously hypertensive rats. Naunyn Schmiedebergs Arch Pharmacol 319 [Suppl]:R48

Rascher W, Lang RE, Fink B, Ganten D, Unger T, Gross F (1982b) Reduced synthesis of (arginine)-vasopressin in spontaneously hypertensive rats. Clin Sci 63:117s–119s

Rascher W, Lang RE, Unger T, Ganten D, Gross F (1982c) Vasopressin in brain of spontaneously hypertensive rats. Am J Physiol 242:H496–H499

Rascher W, Lang RE, Weidemann E, Ganten D, Unger T, Gross F (1982d) Evidence against a role of circulating vasopressin in the pathogenesis of spontaneous hypertension in rats. In: Rascher W, Clough D, Ganten D (eds) Hypertensive mechanisms, Schattauer, Stuttgart, pp 574–576

Rascher W, Lang RE, Ganten D, Meffle H, Taubitz M, Unger T, Gross F (1983) Vasopressin in deoxycorticosterone acetate hypertension of rats: a hemodynamic analysis. J Cardiovasc Pharmacol 5:418–425

Rhodes CH, Morrell JI, Pfaff DW (1981) Immunohistochemical analysis of magnocellular elements in rat hypothalamus: distribution and numbers of cells containing neurophysin, oxytocin and vasopressin. J Comp Neurol 198:45–64

Richardson PDA, Withrington PG (1978) The effects of intraarterial and intraportal injections of vasopressin on the simultaneously perfused hepatic arterial and portal venous vascular beds of the dog. Circ Res 43:496–503

Robertson GL (1977) The regulation of vasopressin function in health and disease. Recent Prog Horm Res 33:333–385

Robertson GL, Athar S (1976) The interaction of blood osmolality and blood volume in regulating plasma vasopressin in man. J Endocrinol Metab 42:613–620

Robertson GL, Mahr EA, Athar S, Sinha T (1973) Development and clinical application of a new method for the radioimmunoassay of arginine-vasopressin in human plasma. J Clin Invest 52:2340–2352

Robertson GL, Aycinana P, Zerbe RL (1982) Neurogenic disorders of osmoregulation. Am J Med 72:339–353

Rocha e Silva M Jr, Rosenberg M (1969) The release of vasopressin in response to hemorrhage and its role in the mechanism of blood pressure regulation. J Physiol (Lond) 202:533–557

Rosenbloom AA, Fisher DA (1975) Radioimmunoassayable AVT and AVP in adult mammalian brain tissue: Comparison of normal and Brattleboro rat. Neuroendocrinology 17:354–361

Rossier J, Pittman Q, Bloom F, Guillemin R (1980) Distribution of opioid peptides in the pituitary: a new hypothalamic-pars nervosa encephalinergic pathway. Fed Proc 39:2555–2560

Rossor MN, Iversen LL, Hawthorn J, Ang VTY, Jenkins JS (1981) Extrahypothalamic vasopressin in human brain. Brain Res 214:349–355

Saavedra J, Grobecker H, Axelrod J (1978) Changes in central catecholaminergic neurons in spontaneously (genetic) hypertensive rats. Circ Res 42:529–534

Sachs H, Takabakate Y (1964) Evidence for a precursor in vasopressin biosynthesis. Endocrinology 75:943–948

Sachs H, Fawcett P, Takabatake Y, Portanova R (1969) Biosynthesis and release of vasopressin and neurophysin. Recent Prog Horm Res 25:447–491

Saito T, Yajima Y (1982) Development of DOCA-salt hypertension in the Brattleboro rat. Ann NY Acad Sci 394:309–318

Sawyer WH, Manning M (1973) Synthetic analogues of oxytocin and vasopressin. Annu Rev Pharmacol 13:5–17
Sawyer WH, Acosta M, Manning M (1974) Structural changes in the arginine vasopressin molecule that prolong its antidiuretic action. Endocrinology 95:140–149
Sawyer WH, Grzonka Z, Manning M (1981 a) Neurohypophyseal peptides: Design of tissue-specific agonists and antagonists. Mol Cell Endocrinol 22:117–134
Sawyer WH, Pang PKT, Seto J, McEnroe M, Lammek B, Manning M (1981 b) Vasopressin analogues that antagonize antidiuretic responses by rats to the antidiuretic hormone. Science 212:49–51
Schmid PG, Abboud FM, Wendling MG, Ramberg ES, Mark AL, Heistad DD, Eckstein JW (1974) Regional vascular effects of vasopressin: plasma levels and circulatory responses. Am J Physiol 227:998–1004
Schömig A, Dietz R, Rascher W, Lüth JB, Mann JFE, Schmidt M, Weber J (1978) Sympathetic vascular tone in spontaneous hypertension of rats. Klin Wochenschr 56 [Suppl I]:131–138
Schrier RW, Berl T, Anderson RJ (1979) Osmotic and nonosmotic release of vasopressin release. Am J Physiol 236:F321–F332
Schulz H, Du Vigneaud V (1966) Synthesis of (1-L-penicillamine) oxytocin, (1-D-penicillamine)oxytocin and (1-deaminopenicillamine)oxytocin, potent inhibitors of the oxytocin response to oxytocin. J Med Chem 9:647–650
Schwartz J, Reid IA (1981) Effect of vasopressin blockade on blood pressure regulation during hemorrhage in conscious dogs. Endocrinology 109:1778–1780
Share L (1961) Acute reduction in extracellular fluid volume and the concentration of antidiuretic hormone in blood. Endocrinology 69:925–933
Share L (1974) Blood pressure, blood volume and the release of vasopressin. In: Greep RO, Astwood EB (eds) The pituitary gland and its neuroendocrine control, Vol IV, Part I. Am Physiol Soc, Washington DC, pp 243–255
Share L, Levy MN (1966) Carotid sinus pulse pressure, a determinant of plasma antidiuretic hormone concentration. Am J Physiol 211:721–724
Share L, Crofton JT, Rockhold RW, Rapp JP (1982) Vasopressin secretion and responses to captopril and a vasopressin antagonist in the Dahl rat. In: Rascher W, Clough D, Ganten D (eds) Hypertensive mechanisms. Schattauer, Stuttgart, pp 577–580
Skowsky WR, Rosenbloom AA, Fischer DF (1974) Radioimmunoassay of arginine vasopressin: development and application. J Clin Endocrinol Metab 38:278–287
Sladek CD, Joynt RJ (1979 a) Characterization of cholinergic control of vasopressin release by the organ-cultured rat hypothalamo-neurohypophyseal system. Endocrinology 104:659–663
Sladek CD, Joynt RJ (1979 b) Cholinergic involvement in osmotic control of vasopressin release by the organ-cultured rat hypothalamo-neurohypophyseal system. Endocrinology 105:367–371
Smith MJ, Cowley AW, Guyton AC, Manning RD (1979) Acute and chronic effects of vasopressin on blood pressure, electrolytes and fluid volumes. Am J Physiol 237:F232–240
Sofroniew MV, Schrell U (1981) Evidence of a direct projection from oxytocin and vasopressin neurons in the hypothalamic paraventricular nucleus to the medulla oblongata. Immunohistochemical visulization of both the horseradish peroxidase transported and the peptide produced by the same neurons. Neurosci Lett 22:211–217
Sofroniew MV, Weindl A (1978 a) Extrahypothalamic neurophysin-containing perikarya, fiber pathway and fiber cluster in the rat brain. Endocrinology 102:334–337
Sofroniew MV, Weindl A (1978 b) Projections from the parvocellular vasopressin- and neurophysin-containing neurons of the suprachiasmatic nucleus. Am J Anat 153:391–401
Starling EH, Verney EB (1924) The secretion of urine as studied on the isolated kidney. Proc R Soc, Lond [Biol] 97:321–363
Swanson LW (1977) Immunohistochemical evidence for a neurophysin containing autonomic pathway in the paraventricular nucleus of the hypothalamus. Brain Res 128:346–353

Szczepanska-Sadowska E (1973) Hemodynamic effects of a moderate increase of the plasma vasopressin level in conscious dogs. Pfluegers Arch 338:313–322
Szczepanska-Sadowska, E, Sobocinska J, Sadowski B (1982) Central dipsogenic effect of vasopressin. Am J Physiol 242:R372–R379
Tanaka M, De Kloet ER, De Wied D, Versteeg DHG (1977 a) Arginine-vasopressin affects catecholamine metabolism in specific brain nuclei. Life Sci 20:1799–1808
Tanaka M, Versteeg DHG, De Wied D (1977 b) Regional effects of vasopressin on rat brain catecholamine metabolism. Neurosci Lett 4:321–325
Thames MD, Schmid PG (1979) Cardiopulmonary receptors with vagal afferents tonically inhibit ADH release in the dog. Am J Physiol 237:H299–H304
Thames MD, Schmid PG (1981) Interaction between carotid and cardiopulmonary baroreflexes in control of plasma ADH. Am J Physiol 241:H431–H434
Thorn NA, Torp-Pedersen, Treiman M, Dartt DA, Worm-Petersen S (1980) Influence of calcium on release of neurohypophyseal hormones. In: Wuttke W et al. (eds) Brain and pituitary peptides. Karger, Basel, pp 118–124
Turner RA, Pierce JG, Du Vigneaud V (1951) The purification and amino acid content of vasopressin preparations. J Biol Chem 191:21–28
Valtin H, Sawyer WH, Sokol HW (1965) Neurohypophyseal principles in rats homozygous and heterozygous for hypothalamic diabetes insipidus (Brattleboro rats). Endocrinology 77:701–706
Vandesande F, Dierickx K, De Mey J (1975) Identification of vasopressin-neurophysin neurons of the rat suprachiasmatic nuclei. Cell Tissue Res 156:377–380
Varma S, Bhuwaneshwar PJ, Bhargava KP (1969) Mechanism of vasopressin-induced bradycardia in dogs. Circ Res 24:787–792
Vávra I, Machová A, Holeček V, Cort JH, Zaoral M, Šorm F (1968) Effect of a synthetic analogue of vasopressin in animals and in patients with diabetes insipidus. Lancet I:948–952
Veale WL, Kasting NW, Cooper KE (1981) Arginine vasopressin and endogenous antipyresis: evidence and significance. Fed Proc 40:2750–2753
Versteeg D, Palkovits M, Van der Gugten J, Wijnen H, Smeets G, De Jong W (1976) Catecholamine content of individual brain regions of spontaneously hypertensive rats (SH-rats). Brain Res 112:429–434
Vilhardt J, Hedquist P (1970) A possible role of prostaglandin E_2 in the regulation of vasopressin secretion in rats. Life Sci 9:825–830
Von den Velden R (1913) Die Nierenwirkung von Hypophysenextrakten beim Menschen. Berlin Klin Wochenschr 45:2083–2086
Wagner HN, Braunwald E (1956) The pressor effect of the antidiuretic principle of the posterior pituitary in orthostatic hypotension. J Clin Invest 35:1412–1418
Walter R (1975) Neurophysins: carriers of peptide hormones. Ann NY Acad Sci 248:1–512
Weindl A, Sofroniew MV (1978) The functional morphology of vascular and neuronal efferent connections of neuroendocrine systems. Arzneimittelforsch 28(II):1264–1268
Weinstein H, Berne RM, Sachs H (1960) Vasopressin in blood: Effect of hemorrhage. Endocrinology 66:712–718
Weitzman RE, Fisher DA (1977) Log linear relationship between plasma arginine vasopressin and plasma osmolality. Am J Physiol 233:E37–E40
Weitzman RE, Reviczky A, Oddie TH, Fisher DA (1980) Effect of osmolality on arginine vasopressin and renin release after hemorrhage. Am J Physiol 238:E62–E68
Wijnen HJLM, Spierenburg HA, De Kloet ER, De Jong W, Versteeg DHG (1980) Decrease in noradrenergic activity in hypothalamic nuclei during the development of spontaneous hypertension. Brain Res 184:152–162
Wisgerhof MV, Northrup TE, Heubelein DM, Brown RD, Dousa TP (1977) Role of antidiuretic hormone in the elevation of blood pressure caused by deoxycorticosterone and NaCl. Fed Proc 36:491 abstract
Woods RL, Johnston CI (1982) Role of vasopressin in hypertension: studies using the Brattleboro rat. Am J Physiol 242:F727–F732
Woods RL, Abrahams JM, Kincaid-Smith P, Johnston CI (1983) Malignant hypertension in Brattleboro (vasopressin-deficient) rats. J Hypertens 1:37–43

Yamamoto M, Share L, Shade RE (1976) Vasopressin release during ventriculo-cisternal perfusion with prostaglandin E_2 in the dog. J Endocrinol (Lond) 71:325–331
Yates FE, Russell DM, Maran JW (1971) Brain adrenohypophyseal communication in mammals. Annu Rev Physiol 33:393–444
Young DB, Pan YJ, Guyton AC (1977) Control of extracellular sodium concentration by antidiuretic hormone-thirst feedback mechanism. Am J Physiol 232:R145–R149
Zerbe RL, Feuerstein G, Meyer DK, Kopin IJ (1982) Cardiovascular, sympathetic, and renin-angiotensin system responses to hemorrhage in vasopressin-deficient rats. Endocrinology 111:608–613
Zimmerman EA (1981) The organization of oxytocin and vasopressin pathways. In: Martin JB, Reichlin S, Bick KL (eds) Neurosecretion and brain peptides. Raven, New York, pp 63–75
Zimmerman EA, Stillman MA, Recht LD, Antunes JL, Carmel PW (1977) Vasopressin and corticotropin-releasing factor: an axonal pathway to portal capillaries in the externa zona of the median eminence containing vasopressin and its interaction with adrenal corticoids. Ann NY Acad Sci 297:405–419

Neuroanatomical Pathways Related to Vasopressin

A. WEINDL and M. SOFRONIEW

Contents

Neurologische Klinik und Poliklinik der Technischen Universität, Möhlstraße 28, D-8000 München 80

Current Topics in Neuroendocrinology, Vol. 4

1 Introduction

Vasopressin and oxytocin are hormones of the neurohypophysis (the posterior pituitary"gland") best known for their peripheral endocrine effects: regulation of antidiuresis and blood pressure, contraction of smooth muscles of the uterus during labour and of the mammary gland during milk ejection respectively. Additional actions on various peripheral organs have been described, e.g., blood clotting (Mannucci et al. 1975), and liver metabolism (Keppens and De Wulf 1979; Martin and Baverel 1981).

A number of central actions of vasopressin and oxytocin have also become known, including neurotropic effects, such as central regulation of blood pressure, cardiovascular regulation, thermoregulation, and behavioral effects in certain test models (see Table 4).

These observations have raised the question of whether the neurosecretory neurons of the hypothalamus which produce the nonapeptides vasopressin and oxytocin, together with larger carrier peptides of still unknown function, the neurophysins, have the functional capacity to deliver their products into the bloodstream from the neurohypophysis, a neurohemal storage organ lacking a blood-brain barrier, as well as to transport them within neural pathways to dis-

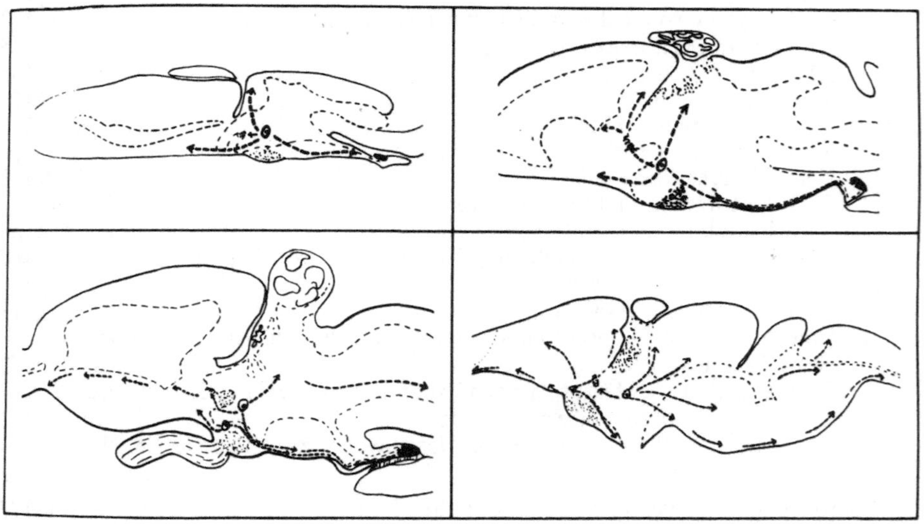

Fig. 1. Early demonstration by Legait and Legait (1957) of extrahypothalamic neurosecretory pathways in the amphibian brain (*Triturus alpestris*)

tinct groups of target neurons within the brain and spinal cord. Early reports on centrally projecting fibers from hypothalamic neurosecretory neurons visualized by neurosecretory staining techniques (Fig. 1) indicated the possibility of a dual functional organization of these neurons (Barry 1956, 1961; Legait 1957; Legait and Legait 1957; Sterba 1974). More substantial evidence for fiber projections to extrahypophyseal neural target areas was derived mainly from the application of immunohistochemical techniques using specific antisera.

This chapter will provide a detailed account of the findings of central connections of neurons producing vasopressin. Other neurohypophyseal peptides, especially oxytocin, will also be discussed in this context. The subject has been reviewed previously by Sofroniew and Weindl (1981), Sofroniew (1982, 1983 a, b), Buijs (1983), Kozlowski et al. (1983), and Silverman and Zimmerman (1983).

2 Overview of Methodological Approaches

Most of the data on the distribution of neurohypophyseal peptides in nerve cell bodies and their processes to hypophyseal and extrahypophyseal vascular target organs and extrahypothalamic neural target areas are derived from the application of immunohistochemical staining methods (immunofluorescence technique of Coons 1956, 1958; unlabeled antibody-enzyme immunoperoxidase technique of Sternberger et al. 1970) on chemically fixed frozen cryostat sections, unembedded vibratome sections, or paraffin sections.

The combination of these immunohistochemical methods with retrograde or anterograde tracing methods has further contributed to the recognition of peptide-containing pathways (Sawchenko and Swanson 1981; Sofroniew and Schrell 1981).

Precise morphological demonstration of the cellular products deriving from larger precursor molecules in neuronal cell bodies has recently become possible by hybridization histochemistry (Coghlan et al. 1984).

Quantitative radioimmunological measurements of neurohypophyseal peptides in microdissected or punched out (Palkovits 1978, 1979) target areas are important to confirm and substantiate the immunohistochemical findings (George and Jacobowitz 1975; Dogterom et al. 1978 a; Glick and Brownstein 1980; Hawthorn et al. 1980; Rossor et al. 1981; Palkovits 1982; Jenkins et al. 1984).

3 The Concept of Neurosecretion and the Chemical Identification of Neurohypophyseal Peptides

A major breakthrough in the understanding of how the brain controls the secretory activity of the posterior pituitary was the observation by Bargmann (1949) of the Gomori-stained (Gomori 1941) neurosecretory material within magnocellular neurons of the dog hypothalamic supraoptic and paraventricular nuclei, as well as within their beaded processes, which terminate at the capillaries of the neural lobe.

The first observations of secretory products in nerve cells, leading to the evolution of the concept of neurosecretion, date back to reports by Speidel (1917), Scharrer (1934a, b), Scharrer and Scharrer (1939), and Gaupp (1944).

By biochemical analysis the neurohypophyseal hormones vasopressin and oxytocin were shown to be nonapeptides with a ring structure (Du Vigneaud 1954). The accompanying "carrier" peptides of vasopressin and oxytocin are large molecules consisting of 90 amino acids (MW 10000) (Acher et al. 1955). Using recombinant cDNA cloning techniques, Land et al. (1982, 1983) have characterized the precursor molecules of vasopressin and oxytocin and their accompanying neurophysins. Arginine-vasopressin derives from a 166-amino-acid precursor molecule, a preprohormone (MW 21000) with a glycosylated side chain. It contains a 19-amino-acid signal peptide followed by arginine-vasopressin, connected to neurophysin II by a Gly-Lys-Arg sequence. The C-terminus contains a naturally occurring glycopolypeptide of 39 amino acids, separated from neurophysin II by a single Arg residue. Oxytocin and neurophysin I derive from a preprohormone (MW 16500) without a glycopolypeptide. Partial characterizations of the precursor molecules had been achieved using conventional techniques by Pickering (1976) as well as by Gainer et al. (1977).

4 The Distribution of Neurohypophyseal Peptides in Magnocellular and Parvocellular Neurons

4.1 Neurosecretory Staining and Immunohistochemistry

Bargmann's description of the neurosecretory neuron in 1949 elicited extensive activity concerning the light- and electron-microscopic evaluation of the hypothalamoneurohypophyseal system in many representatives of all vertebrate classes under normal and experimental conditions, as well as of comparable neurosecretory systems in invertebrates (Scharrer 1967). Some of these studies showed that neurosecretory fibers from the paraventricular nucleus terminated additionally at portal capillaries of the external zone of the median eminence, implicating a role in the control of adenohypophyseal function. Furthermore, neurosecretory fibers were found in other neurohemal target areas lacking a blood-brain barrier (see Table 2 and Fig. 2): the organum vasculosum of the lamina terminalis (OVLT; supraoptic crest, preoptic organ), subfornical organ, and pineal body. In some of these so-called circumventricular organs neurosecretory terminals are in contact with fenestrated capillaries which are permeable for peptides and proteins (Weindl 1973, 1983), thus permitting the delivery of peptide neurohormones into the blood stream.

Several authors using neurosecretory staining techniques found beaded fibers containing neurosecretory material which were not in contact with the capillaries of the neural lobe or of other circumventricular organs. Particularly in submammalian vertebrates, numerous fibers containing neurosecretory material were

Fig. 2. Circumventricular organs of the human brain (*AP*, area postrema; *ME*, median eminence; *NH*, neural lobe of hypophysis; *OVLT*, organum vasculosum of lamina terminalis; *PI*, pineal body; *SCO*, subcommissural organ; *SFO*, subfornical organ; *CP*, choroid plexus). With the exception of the SCO, all circumventricular organs are located outside the blood-brain barrier and are permeable to peptides

found which projected from the hypothalamus to various telencephalic, diencephalic, mesencephalic, and rhombencephalic areas (Fig. 1) (Ananthanrazanan 1955; Barry 1956, 1961; Legait 1957; Legait and Legait 1957; Sterba 1974).

The development of immunohistochemical techniques provided the possibility of identifying the specific peptides contained in neurosecretory cells and their processes.

Neurophysin was first demonstrated by the immunofluorescence technique in the hypothalamoneurohypophyseal system of the dog (Alvarez-Buylla et al. 1970). Soon a wealth of publications appeared demonstrating the precise distribution of vasopressin, oxytocin, and their related neurophysins in mammals (for reviews see Watkins 1975; Dierickx 1980).

The distribution of the phylon-specific analogues of vasopressin and oxytocin and their related neurophysins in birds, reptiles, amphibia and fish is demonstrated in Fig. 3.

Grimmelikhuijzen et al. (1982) observed in the hydra nerve cells containing oxytocin and vasopressin-like material. In arthropods, vasopressin-like, oxy-

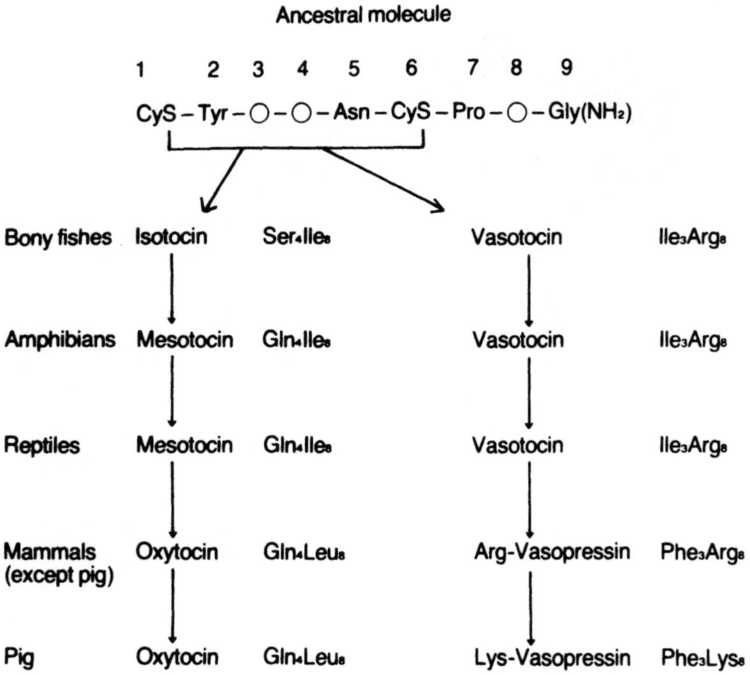

Fig. 3. Survey of the neurohypophyseal peptide hormones in vertebrate evolution. Hypothetical scheme compiled by Acher (1980). One gene duplication and a series of subsequent single substitutions in position 3, 4, or 8 produce two molecular lines. The substituted amino acids and their position in the hormone are listed to the right of each hormone

tocin-like, and neurophysin-like immunoreactive nerve cells were described in the suboesophageal ganglion which had previously been classified as neurosecretory cells by means of neurosecretory staining techniques and by electron-microscopic demonstration of dense-core elementary granules (Girardie and Remy 1980; Remy and Girardie 1980). Similar findings in the Colorado potato beetle were reported by Veenstra et al. (1984). (For a review of peptides related to vasopressin in invertebrates, see Mühlethaler et al. 1984).

The findings of extrahypothalamic projections of neurosecretory neurons to neural target areas (Barry 1956, 1961; Legait 1957; Legait and Legait 1957; Sterba 1974) were confirmed by observations of special patterns of extrahypothalamic fiber projections containing neurohypophyseal peptides on the basis of a great number of immunohistochemical observations in many species representing all vertebrate classes.

4.2 Vasopressin- and Neurophysin-Producing Perikarya

4.2.1 Magnocellular Perikarya of the Hypothalamus

The use of immunohistochemical techniques has essentially confirmed and extended previous observations obtained by neurosecretory staining methods on the magnocellular hypothalamoneurohypophyseal system (Dierickx and Vandesande 1979 a). Vasopressin or oxytoxin and their associated neurophysins are produced in magnocellular perikarya of the supraoptic and paraventricular nuclei (Figs. 4 a–f, 5) and in several groups of accessory neurosecretory perikarya (for review, see Watkins 1975; Dierickx 1980; Silverman and Zimmerman 1983). The 39-amino-acid precursor to arginine-vasopressin has been demonstrated immunohistochemically within the magnocellular hypothalamoneurohypophyseal system (Watson et al. 1982 b) and in the suprachiasmatic nucleus (Lu et al. 1982).

4.2.1.1 The Paraventricular and Supraoptic Nuclei

The *paraventricular nucleus* in the rat has been subdivided into several nuclear groups (Swanson and Kuypers 1980; Armstrong et al. 1980). Swanson and Kuypers (1980) described eight cytoarchitectonic subdivisions of the paraventricular nucleus, three magnocellular (anterior, posterior, and medial) and five parvocellular (anterior, medial, lateral, dorsal, and periventricular) (Fig. 6). Armstrong et al. (1980) used criteria of Nissl staining, Golgi impregnation, immunohistochemical neurophysin staining, and retrograde transport of horseradish peroxidase and described the following subdivisions (Fig. 7): The anterior commissural nucleus (1) contains many magnocellular neurophysin perikarya projecting to the neurohypophysis; one or two dendrites are directed toward the third ventricle. The medial (2) and lateral (3) magnocellular portions of the paraventricular nucleus have similar but slightly larger and more closely packed cells. Cells of the dorsomedial cap (4) and extreme posterior portion (5) of the paraventricular nucleus often contain neurophysin and are connected to the brain stem and spinal cord. Large cells of the extreme posterior subnucleus are more fusiform, possess more dendrites, and are loosely packed.

Investigation of the intrinsic organization of the paraventricular nucleus of the rat (Van den Pol 1982) revealed differences in the synaptic connectivity between magnocellular and parvocellular divisions. Swanson et al. (1981) have summarized the major cell groups of the paraventricular nucleus in a schematic diagram (Fig. 6).

According to Swanson and Sawchenko (1980), most neurons of the magnocellular division contain either vasopressin or oxytocin and project to the neural lobe. At the ultrastructural level, vasopressin was found in neurosecretory granules of perikarya of the posterior magnocellular division of the paraventricular nucleus (Piekut 1983). Neurophysin perikarya of the posterior subnucleus of the paraventricular nucleus were found to contain estrogen receptors (Rhodes et al. 1981). Parvocellular perikarya project to the median eminence or brain stem and spinal cord. Magnocellular vasopressin neurons are innervated preferentially by noradrenergic fibers, parvocellular neurons by adrenergic fibers from the brain stem, dorsal vagal complex, and parabrachial nucleus (Swanson et al. 1981; Sawchenko and Swanson 1982 b). Some oxytocin neurons of the paraventricular nucleus also contain tyrosine hydroxylase and appear to be catecholaminergic (Schimchowitsch et al. 1983).

The distribution of magnocellular vasopressin/neurophysin and oxytocin/neurophysin perikarya in the *supraoptic nucleus* is less diversified than in the

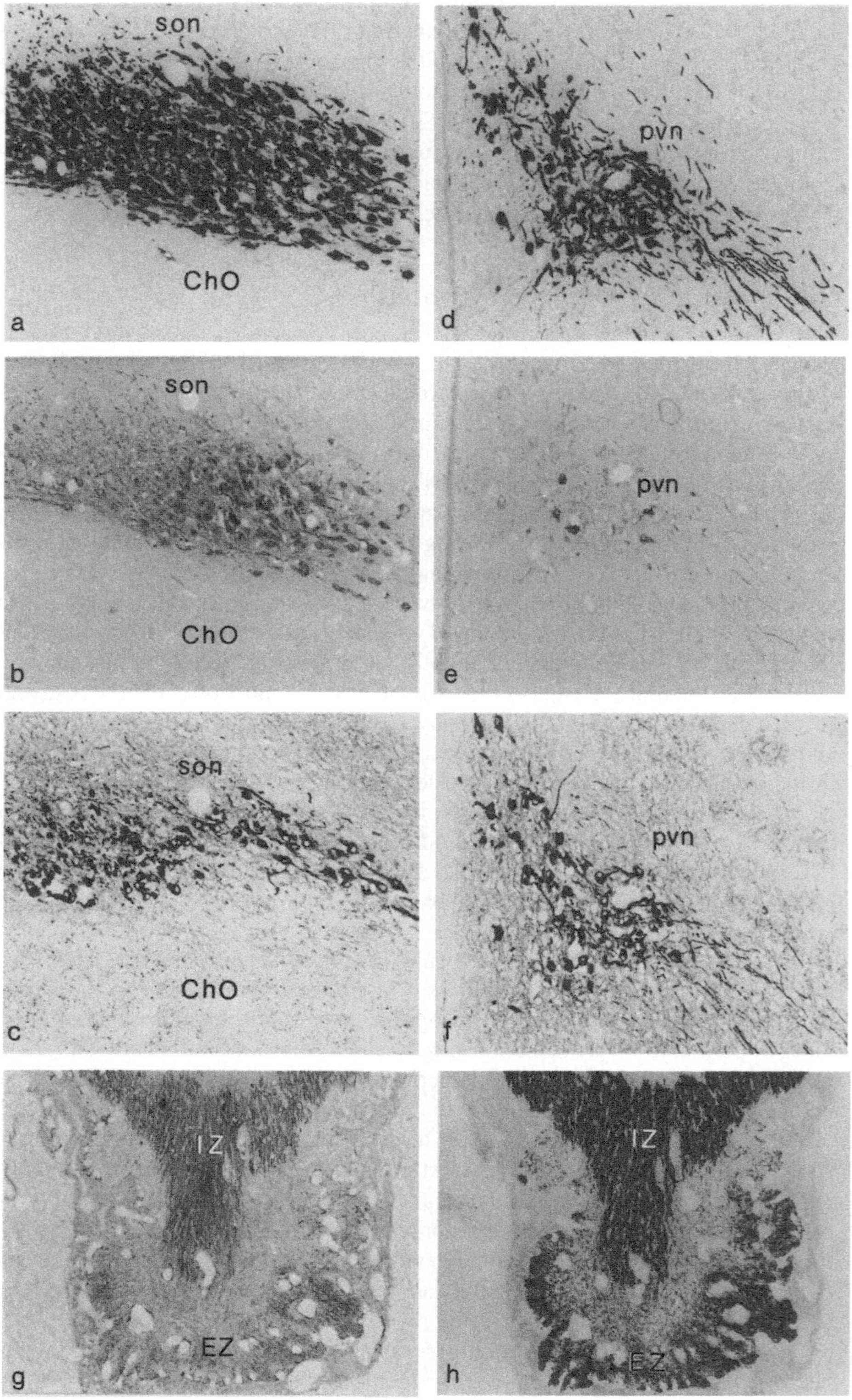
son
ChO
a
pvn
d
son
ChO
b
pvn
e
son
ChO
c
pvn
f
IZ
EZ
g
IZ
EZ
h

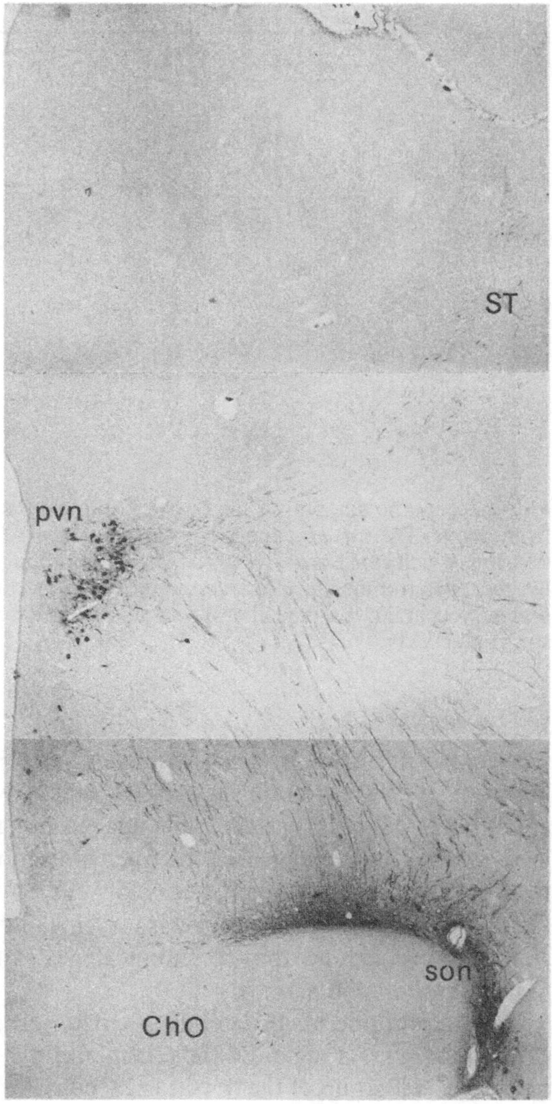

Fig. 5. *Tupaia.* Vasopressin immunoperoxidase reaction. Frontal section through diencephalon. Vasopressin perikarya in supraoptic (*son*) and paraventricular nucleus (*pvn*) and internuclear zone. Vasopressin fibers project ventrolaterally (mainly to pituitary), laterally, and dorsolaterally to the stria terminalis (*ST*). *ChO*, chiasma opticum. 32x

---

◀

Fig. 4 a–h. *Sanguinus fuscicollis.* **a, d, h** Neurophysin, **b, e, g** vasopressin, and *c, f* oxytocin immunoperoxidase reaction. Neighboring frontal sections through **a–c** supraoptic (*son*) and **d–f** paraventricular nucleus (*pvn*) and **g, h** infundibulum. A regional pattern of distribution of vasopressin and oxytocin perikarya is more pronounced in the pvn. In the infundibulum, vasopressin and neurophysin fibers terminate at portal capillaries of the external zone (*EZ*). The internal zone (*IZ*) contains vasopressin and neurophysin fibers of the hypothalamoneurohypophyseal tract. *ChO*, chiasma opticum. **a–h,** 80x

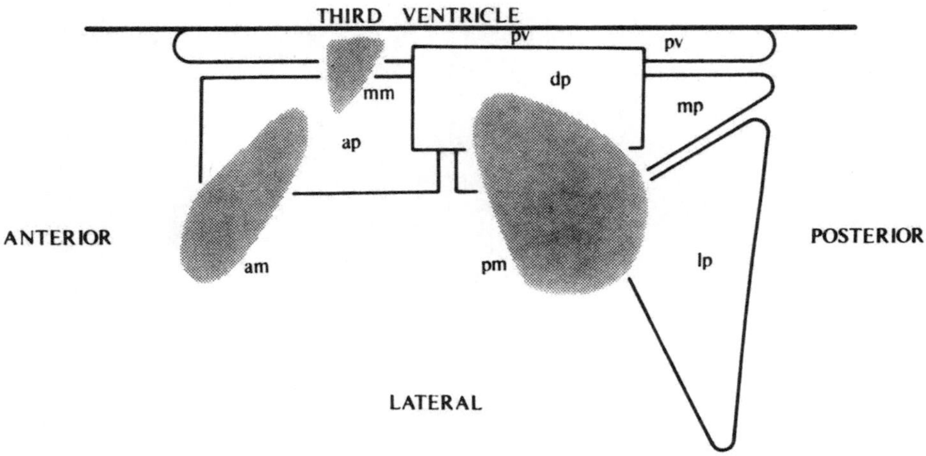

Fig. 6. Schematic diagram of major cell groups in the paraventricular nucleus, as viewed from above. The three magnocellular groups (*stippled*) are embedded in the parvocellular division, which consists of five parts: *am*, anterior magnocellular; *mm*, medial magnocellular; *pm*, posterior magnocellular; *ap*, anterior parvocellular; *dp*, dorsal parvocellular; *lp*, lateral parvocellular; *mp*, medial parvocellular; *pv*, periventricular parvocellular. From Swanson et al. (1981)

paraventricular nucleus. Their dendrites are oriented toward the external glial lamina of the ventral brain surface (Dyball and Kemplay 1982; W. E. Armstrong et al. 1982). Many axons leave the nucleus dorsally and turn toward the neural lobe. Only a few reports describe fiber projections from the supraoptic nucleus to extrahypothalamic target sites (Buijs 1983; Alonso et al. 1984). Mason et al. (1984), using horseradish peroxidase tracing and electrical recording, found that supraoptic nucleus neurons communicate via axon collaterals with other neurons in the lateral hypothalamus.

The supraoptic nucleus receives a rich noradrenergic input (Swanson et al. (1981) from the A1 region of the ventral medulla oblongata and the A2 catecholaminergic cell group of the dorsal vagal complex in the medulla oblongata (Sawchenko and Swanson 1982 b; Sladek and Sladek 1983). Choline acetyltransferase-positive perikarya projecting to the supraoptic nucleus were found dorsal and lateral to the supraoptic nucleus (Mason et al. 1983).

Afferent Projections to the Paraventricular and Supraoptic Nuclei. Afferent connections to the paraventricular nucleus have been traced by several authors (Ricardo and Koh 1978; McKellar and Loewy 1981; Silverman et al. 1981 b; Tribollet and Dreifuss 1981). Sawchenko and Swanson (1983) and Swanson and Sawchenko (1983) have reviewed the afferent connections to the paraventricular and supraoptic nucleus. Noradrenergic terminals in both nuclei derive from the A1 cell group in the ventrolateral medulla oblongata and the A6 cell group in the locus coeruleus. Only the A1 group of the ventral medulla oblongata projects substantially to the supraoptic nucleus and the magnocellular division of the

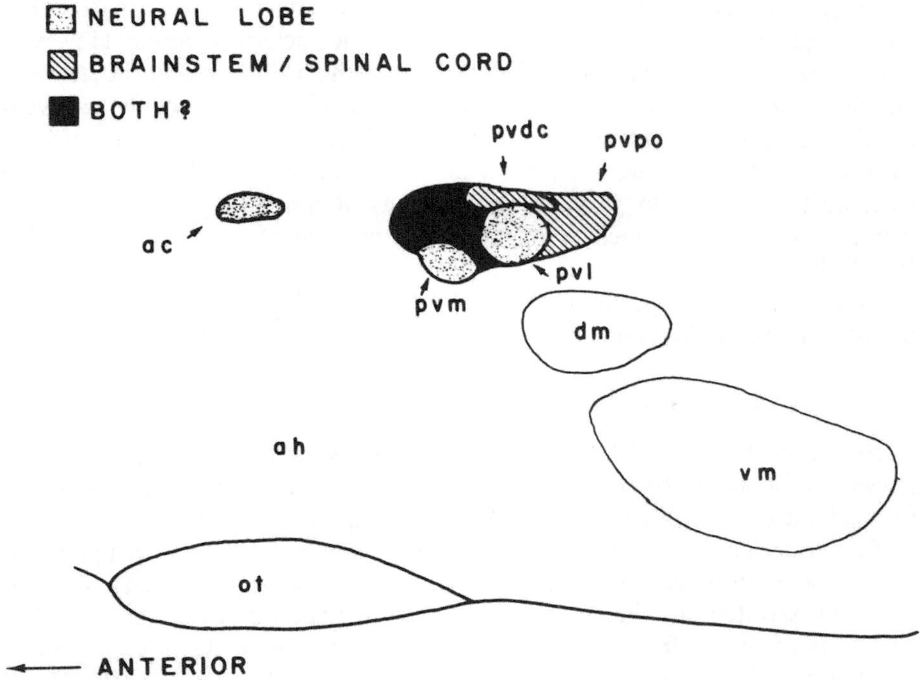

Fig. 7. Schematic representation of the anterior commissural (*ac*) and paraventricular nuclei in sagittal section, depicting their relationship with the neurohypophysis and brain stem/spinal cord. The nuclei have been divided on the basis of the efferent projections to either brain stem/spinal cord or neural lobe. *Both* indicates that these areas contain cells which project to both places, but whether individual cells are responsible for this dual innervation is unknown. *ah,* anterior hypothalamic nucleus; *dm,* dorsomedial hypothalamic nucleus; *ot,* optic tract; *pvdc,* dorsomedial cap of the paraventricular nucleus; *pvm,* medial magnocellular paraventricular nucleus; *pvpo,* posterior part of the paraventricular nucleus; *pvl,* lateral magnocellular paraventricular nucleus; *vm,* ventromedial hypothalamic nucleus. From Armstrong et al. (1980)

paraventricular nucleus. The A1 group receives afferent connections from the solitary tract nucleus, the major recipient of first order visceral afferents. Visceral afferents from atrial stretch receptors, carotid sinus baroreceptors, and carotid body chemoreceptors are conveyed to the solitary tract nucleus, A1 region (or via the parabrachial nucleus) to the magnocellular vasopressin neurons. Other afferents to magnocellular vasopressin neurons derive from the subfornical organ (Miselis 1981; Sawchenko and Swanson 1983) and the median preoptic area (Sawchenko and Swanson 1983).

Serotonin fibers from the dorsal and median raphe nuclei (B7, B8, B9) project mainly to oxytocin neurons in the paraventricular and supraoptic nuclei. Fibers from the arcuate nucleus region containing ACTH 1–39 (Sawchenko et al. 1982) or other neuroactive substances (dopamine, acetylcholine) project to parvocellular neurons of the paraventricular nucleus. Oxytocin neurons receive additional inputs from the median preoptic nucleus, subfornical organ, and dorsomedial hypothalamic nucleus. No substantial fiber projections from the septum, subiculum, and amygdala to the paraventricular and supraoptic nuclei were found. However, fiber projections from the amygdala terminate in the bed nucleus of the stria terminalis, which in turn is connected to the parvocellular division of the paraventricular nucleus and moderately to the magnocellular division. The preoptic and various hypothalamic nuclei, parabrachial nucleus, locus coeruleus, and solitary tract nucleus project to parvocellular neurons of the paraventricular nucleus which in turn are connected to magnocellular neurons of this nucleus (Van den Pol 1982).

According to lesion experiments by Meyer et al. (1980), the supraoptic nucleus receives afferent GABAergic fibers from the nucleus accumbens.

4.2.1.2 Accessory Magnocellular Neurons

According to Peterson (1966), in the rat the following accessory magnocellular nuclei can be identified: rostral periventricular nucleus, anterior and posterior fornical nuclei, nucleus circularis, nucleus of median forebrain bundle, and internuclear zone. The spatial distribution of vasopressin neurons in the magnocellular accessory neurosecretory and paraventricular and supraoptic nuclei of the rat was demonstrated in a three-dimensional reconstruction by Fisher et al. (1979). There are species differences concerning the distribution of the accessory magnocellular nuclei as described in the mouse (Silverman and Zimmerman 1983) and guinea pig (Sofroniew et al. 1979).

Both vasopressin and oxytocin and their associated neurophysins are present in the accessory magnocellular nuclei (Reaves and Hayward 1979). The anterior commissural nucleus contains only oxytocin neurons (Sofroniew and Glasmann 1981).

4.2.1.3 Other Peptides in Hypothalamic Neurosecretory Nuclei and the Question of Coexistence of Peptides

With the detection of a rapidly increasing number of peptides in the central and peripheral nervous system, additional peptides have been localized within magnocellular hypothalamic nuclei, especially in the paraventricular nucleus. So far, over 30 possible neurotransmitters have been found within the paraventricular and supraoptic nuclei. Ten neuroactive substances have been described within cell bodies in the paraventricular nucleus (Sawchenko and Swanson 1983). Some of these peptides were localized in various subgroups (see Sect. 4.2.1.1) of the paraventricular nucleus.

Within the paraventricular nucleus, *vasoactive intestinal polypeptide*-containing perikarya were described by Sims et al. (1980). *Angiotensin II*-immunoreactive perikarya were reported in neurons of the paraventricular and supraoptic nuclei of the rat by Ganten et al. (1978), by Fuxe et al. (1982), by Phillips et al. (1979), by Kilcoyne et al. (1980) and by Zimmerman et al. (1980). Kilcoyne et al. (1980) suggest that angiotensin II may be colocalized with vasopressin in some of the magnocellular perikarya. Zimmerman et al. (1980) report that angiotensin II immunoreactivity is absent in vasopressin neurons of the Brattleboro rat and that angiotensin II immunoreactivity is increased after adrenalectomy. Fuxe et al.

(1982) describe renin-like immunoreactivity in lateral oxytocin neurons of the paraventricular nucleus.

Oxytocin has been shown to coexist with *gastrin/cholecystokinin 8* in some of the rat and bovine magnocellular neurons (Lorén et al. 1979; Beinfeld et al. 1980; Vanderhaeghen et al. 1980; Vanderhaeghen et al. 1981). Kiss et al. (1984) found that magnocellular cholecystokinin (CCK) neurons project to the neural lobe, most parvocellular CCK neurons to the median eminence, and only very few parvocellular CCK neurons to the lower brain stem.

Earlier reports by Sar et al. (1978) and Rossier et al. (1979) in the rat and by Micevych and Elde (1980) in the cat indicated *enkephalin* immunoreactivity in neurons of the supraoptic and paraventricular nuclei. At the electron-microscopic level coexistence of neurophysin and enkephalin was found in the same terminals and even in the same granules (Coulter et al. 1981) of the neural lobe. Martin et al. (1983 b) found Met-5-enkephalin, a proenkephalin-A-derived opioid peptide fragment, in granules of the rat neurohypophysis. Vanderhaeghen et al. (1983) reported a coexistence of proenkephalin, metenkephalin and oxytocin in some perikarya of the dorsal part of the supraoptic nucleus and the peripheral part of the paraventricular nucleus, as well as in their fibers terminating at fenestrated capillaries of the neural lobe.

Coexistence of *dynorphin,* alpha-neoendorphin and leuenkephalin with vasopressin in some magnocellular perikarya and their fibers and terminals in the neural lobe was reported by Watson et al. (1982 a), Whitnall et al. (1983), Weber et al. (1982 a, b), by Weber et al. (1983), and Martin et al. (1983 a). Watson et al. (1982 a) demonstrated that in the Brattleboro rat, dynorphin immunoreactivity persists in magnocellular neurons despite the absence of vasopressin.
Sawchenko et al. (1984) observed in colchicine-treated rats a colocalization of *corticotropin-releasing factor* (CRF) with oxytocin in a discrete anterior part of the magnocellular division of the paraventricular nucleus and in a subset of neurons of the supraoptic nucleus. A small number of neurons in the parvocellular division of the paraventricular nucleus express both CRF and vasopressin immunoreactivity (Roth et al. 1982). After adrenalectomy, CRF and vasopressin immunoreactivity in presumed parvocellular neurosecretory neurons is enhanced. According to Burlet et al. (1983) CRF neurons are found in two distinct systems: (1) the hypothalamoneurohypophyseal system, where CRF is colocalized in oxytocin-neurophysin I neurons, and (2) the hypothalamoinfundibular system.

Furthermore, *neurotensin* immunoreactivity was reported in parvocellular neurons of the paraventricular nucleus (Kahn et al. 1980).

Sawchenko et al. (1984) reported colocalization of *neurotensin* and *CRF* in the parvocellular division of the paraventricular nucleus.

Tager et al. (1980) reported *glucagon* immunoreactivity within neurons of the paraventricular nucleus.

Using immunohistochemical criteria, especially with respect to cross-reactivity of antisera, and other criteria of biochemistry and cell biology very strictly, at present only coexistence of dynorphin with vasopressin and of CCK 8 peptide with oxytocin in some magnocellular neurons appears likely (Sofroniew et al. 1984).

4.2.2 Parvocellular Vasopressin/Neurophysin Perikarya

4.2.2.1 Suprachiasmatic Nucleus

Parvocellular vasopressin- and neurophysin-containing neurons are located in a more rostral, dorsal, and medial portion of the suprachiasmatic nucleus of the rat (Vandesande et al. 1974, 1975; Sofroniew and Weindl 1982) and a great variety of other mammals including man (Sofroniew and Weindl 1980; Ueda et al. 1983). At the ultrastructural level, vasopressin immunoreactivity is present in elementary granules (Van Leeuwen et al. 1978). Vasopressin was identified radioimmunologically in extracts of the suprachiasmatic nucleus by Kok et al. (1983). Vasopressin and neurophysin immunoreactivity is absent in homozygous Brattleboro rats (Sofroniew and Weindl 1978; Weindl and Sofroniew 1980). Other perikarya of the suprachiasmatic nucleus contain additional peptides, including vasoactive intestinal polypeptide (Sims et al. 1980; Card et al. 1981) and somatostatin (Dierickx and Vandesande 1979b).

The suprachiasmatic nucleus receives various afferent fibers (for review see Pickard 1982) including retinohypothalamic fibers terminating in ventral, lateral, and caudal portions (Moore 1973), and serotonin fibers (Fuxe 1965) terminating mainly ventrally. The serotonin fibers originate in the mesencephalic raphe nuclei (Moore et al. 1978). Fibers containing neuropeptide Y, a representative of the pancreatic polypeptide family, originate from the ventral geniculate body and terminate ventrolaterally (Card and Moore 1982; Moore et al. 1984).

4.2.2.2 Additional Parvocellular Vasopressin/ Neurophysin Perikarya

Additional parvocellular perikarya immunoreactive with vasopressin and neurophysin II antisera were observed in rats only after colchicine treatment, which in-

Table 1. Distribution of magnocellular (m) and parvocellular (p) perikarya containing neurohypophyseal peptides

	Vasopressin	Oxytocin	References
Telencephalon			
Nucleus interstitialis striae terminalis	p		Van Leeuwen and Caffé (1983)
Amygdala	p		Caffé and Van Leeuwen (1983)
Diencephalon			
Nucleus paraventricularis	m, p	m, p	Vandesande and Dierickx (1975)
Nucleus supraopticus	m	m	Vandesande and Dierickx (1975)
Accessory neurosecretory nuclei	m	m	Reaves and Hayward (1979)
Nucleus suprachiasmaticus	p		Vandesande et al. (1975)
Nucleus dorsomedialis hypothalami	p		Caffé and Van Leeuwen (1983)
Rhombencephalon			
Locus coeruleus	p		Caffé and Van Leeuwen (1983)

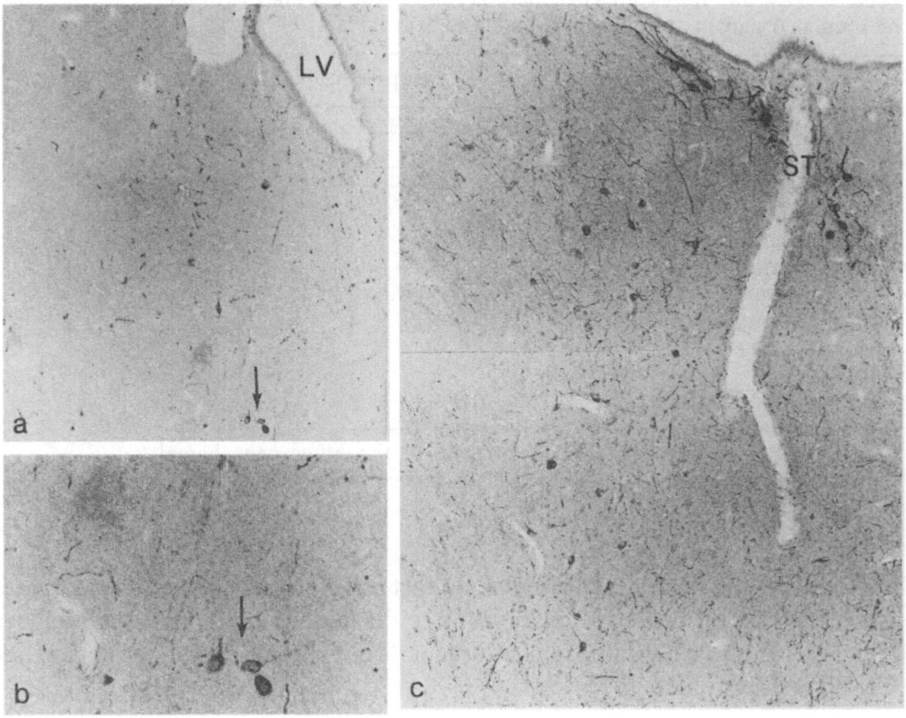

Fig. 8 a–c. *Tupaia*. Neurophysin immunoperoxidase reaction. Frontal sections through bed nucleus of the stria terminalis (*ST*), which contains parvocellular vasopressin/neurophysin perikarya (*arrows*). *LV*, lateral ventricle. **a, c** 80x; **b** detail of **a** (200x)

hibits axoplasmic transport and causes accumulation of secretory products in the perikaryon (see Table 1). Such perikarya were described in the dorsal and ventral portion of the bed nucleus of the stria terminalis (Van Leeuwen and Caffé 1983) (see also Fig. 8 a–c) and in the dorsomedial hypothalamic area, medial amygdaloid nucleus, and locus coeruleus (Caffé and Van Leeuwen 1983). These cells could not be demonstrated in homozygous Brattleboro rats, nor in normal rats after staining with oxytocin and neurophysin I antisera (Van Leeuwen and Caffé 1983).

5 Fiber Projections of Vasopressin/Neurophysin and Oxytocin/Neurophysin Neurons

5.1 Vascular Target Areas

The best-known vascular target area of the classical hypothalamoneurohypopyseal neurosecretory system is the neural lobe. However, there are other vascular target areas where vasopressin/neurophysin and oxytocin/neurophysin fibers ter-

Table 2. Vascular target areas of neurohypophyseal peptides (circumventricular organs) according to Weindl and Sofroniew (1981)

	Vasopressin	Oxytocin
Neural lobe	+++++	+++++
Median eminence		
Zona interna	++++	++++
Zona externa	+++	
Organum vasculosum laminae terminalis		
Zona interna	+	
Zona externa	(+)	
Subfornical organ	+	
Area postrema	−	−
Pineal body	+	(+)
Subcommissural organ[a]	(+)	(+)

[a] In the subcommissural organ, in contrast to the other neurohemal circumventricular organs, the blood-brain barrier is not lacking

minate: these are other neurovascular contact zones lacking a blood-brain barrier which, because of their close relationship to the midline ventricular system, have been termed the circumventricular organs (Table 2 and Fig. 2; for review see Weindl and Sofroniew 1978, 1981).

5.1.1 Neural Lobe

From magnocellular vasopressin/neurophysin or oxytocin/neurophysin perikarya of the supraoptic and paraventricular nuclei and of the accessory magnocellular nuclei, beaded, large-caliber, predominantly unmyelinated fibers project through the internal zone of the median eminence to the neural lobe, where they bilaterally from terminals at permeable fenestrated capillaries. In the paraventricular nucleus, the fibers originate from all parts of the magnocellular division and form scattered cells of the parvocellular division (Swanson and Kuypers 1980). Retrograde labeling studies indicate that the neural lobe receives fibers not only from vasopressin and oxytocin perikarya of the paraventricular and supraoptic nuclei (Sherlock et al. 1975), but also from vasopressin and oxytocin perikarya of the bed nucleus of the stria terminalis, the median and periventricular preoptic nuclei, the medial and lateral preoptic areas, the zona incerta, and the substantia innominata (Kelly and Swanson 1980). The neural lobe is the organ of storage and release of vasopressin, oxytocin, and neurophysin, and contains the highest concentrations of readily releasable peptides within the nervous system. Of the various additional peptides found in neurosecretory neurons and in the neural lobe (Coulter et al. 1981; Martin and Voigt 1981), dynorphin appears to be released with vasopressin, and CCK 8 with oxytocin from some terminals. Van Leeuwen et al. (1983) and Van Leeuwen and De Vries (1983) described synaptoid contacts between enkephalin terminals and pituicytes. They proposed that enkephalin acts on pituicytes and that pituicytes may play an intermediate role in vasopressin and oxytocin release.

5.1.2 Median Eminence

From the parvocellular division of the paraventricular nucleus, vasopressin- and neurophysin-containing fibers project to fenestrated portal capillaries of the external zone of the median eminence, or infundibulum (Fig. 4g, h) (Antunes et al. 1977, Dierickx et al. 1976; Swanson et al. 1980).

Retrograde tracing studies after injection of horseradish peroxidase and true blue into the external zone of the median eminence demonstrated heavy labeling in the periventricular, medial (dorsal half), and anterior parts of the parvocellular division of the paraventricular nucleus (Wiegand and Price 1980; Lechan et al. 1980). Gomori-positive (Wittkowski and Bock 1972) and neurophysin-stained (Silverman and Zimmerman 1975) secretory granules in the external zone of the median eminence are distinctly smaller than in the neural lobe suggesting separate origins of the fibers. Elektrophysiological data from Blume et al. (1978) and Pittman et al. (1978) are in support of a fiber pathway from the paraventricular nucleus to the external zone of the median eminence. After reserpine treatment, vasopressin and neurophysin terminals in the external zone of the median eminence are markedly depleted (Seybold et al. 1981).

The number of vasopressin and neurophysin terminals around portal capillaries of the external zone of the median eminence increases after adrenalectomy. This effect is suppressed by glucocorticoid but not mineralocorticoid substitution (Stillman et al. 1977; Silverman et al. 1981 a; Sofroniew et al. 1977). Adrenalectomy induces increased ^{3}H-cytidine incorporation in paraventricular nucleus neurons (Silverman et al. 1980 a) and sprouting of paraventricular nucleus axons (Silverman and Zimmerman 1982). Immediately after adrenalectomy, vasopressin and CRF immunoreactivity decreases in terminals in the external zone of the median eminence. After 5 days vasopressin and CRF immunoreactivity is increased, vasopressin more than CRF (Bugnon et al. 1983). This suggests that two infundibular systems are involved in ACTH secretion (Kawata et al. 1983 a). High levels of vasopressin and neurophysin were observed in the portal blood of monkeys (Zimmerman et al. 1973 a). After adrenalectomy, vasopressin and neurophysin plasma levels are increased in rats (Seif et al. 1978). These findings suggest that this vasopressinergic pathway may participate in ACTH release (for review see Makara et al. 1981), in addition to CRF neurons projecting from the paraventricular nucleus to the portal vessels. Sawchenko et al. (1984) observed that in this group of neurons vasopressin and CRF are colocalized; however, CRF production in these neurons is only expressed after adrenalectomy.

5.1.3 Organum Vasculosum of Lamina Terminalis

Like the median eminence, the OVLT consists of two zones, the richly vascularized external zone and the internal or parenchymal zone. Camacho and Phillips (1981), using horseradish peroxidase tracing, found neural connections to several hypothalamic nuclei (anterior, median, and lateral preoptic nuclei and ventromedial nucleus) and to extrahypothalamic regions (subfornical organ, central gray, and locus coeruleus). Small-caliber vasopressin and neurophysin fibers are found in the internal zone of the rat but not of primates (Fig. 9a). It appears that

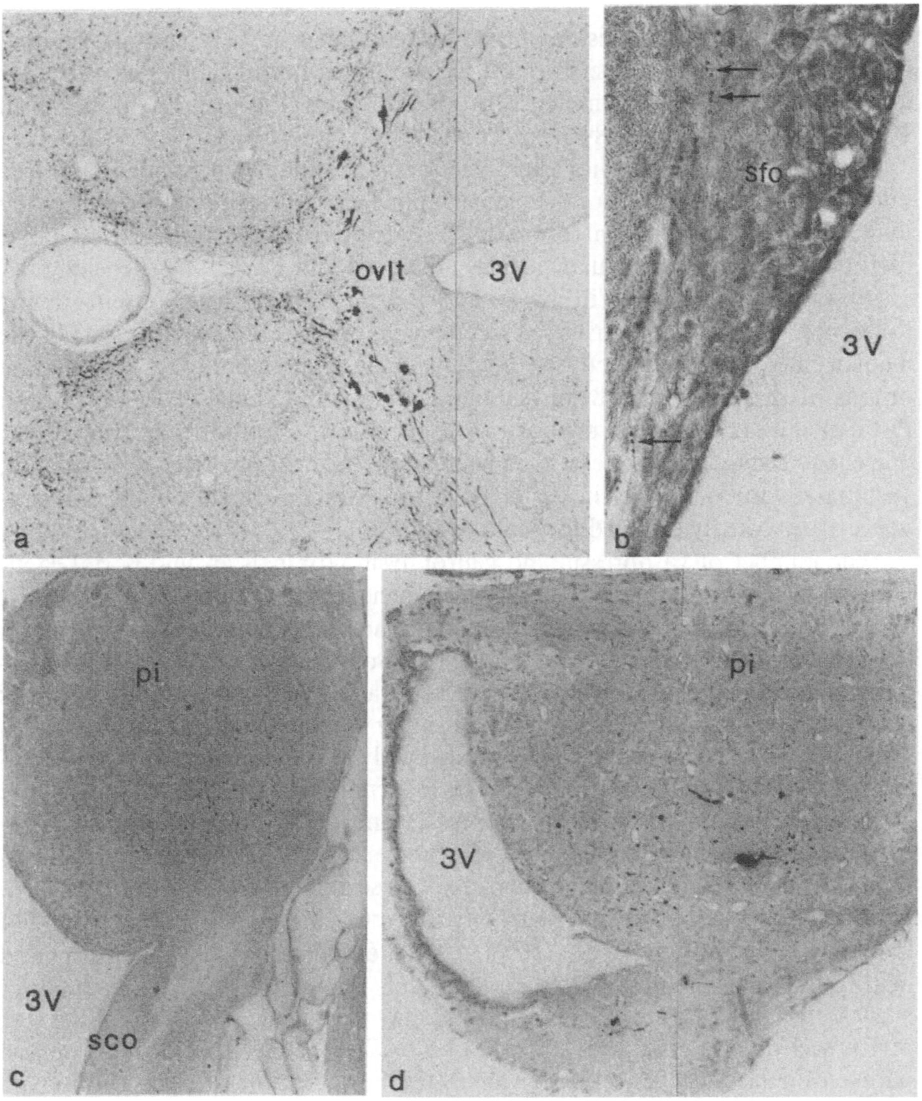

Fig. 9 a–d. *Tupaia.* **a–c** Neurophysin and **d** vasopressin immunoperoxidase reaction. **a** Horizontal, **b, c** sagittal, and **d** frontal sections through organum vasculosum of the **a** lamina terminalis (*ovlt*), **b** subfornical organ (*sfo*), **c, d** pineal (*pi*) and **c** subcommissural organ (*sco*). **a** Neurophysin fibers of the diagonal tract in immediate vicinity of the ovlt; **b** few neurophysin and vasopressin fibers in sfo (*arrows*) and **c, d** pineal. *3V*, 3rd ventricle. **a** 50x; **b** 200x; **c, d** 80x

most of these fibers pass through the organ, because only few fibers are found in contact with permeable fenestrated capillaries of the external zone (Weindl and Sofroniew 1981). Noteworthy is that in the guinea pig, the distribution of magnocellular oxytocin/neurophysin perikarya of the anterior commissural nucleus extends ventrally into the lamina terminalis (Weindl and Sofroniew 1978, 1981).

These perikarya closely surround the OVLT. The perikarya and their dendrites reaching into the OVLT are exposed to molecules which enter the neurohemal zone of the OVLT but do not have access to neurons in other parts of the central nervous system due to the blood-brain barrier. This close relationship may be the structural basis of a receptor mechanism for sensing blood-borne molecules.

5.1.4 Subfornical Organ

In the subfornical organ only scattered vasopressin, oxytocin, and neurophysin fibers (Fig. 9 b) are observed whose origins are not known. Lesion and tracing studies (Miselis 1981, 1982; Renaud et al. 1983) have revealed that afferent connections from the subfornical organ exist to magnocellular neurons in the supraoptic and paraventricular nucleus as well as to other perikarya involved in water regulation and drinking behavior. Radioimmunoassayable vasopressin in the subfornical organ of the rat was increased after water deprivation (Summy-Long et al. 1978).

Neurons in the subfornical organ have been shown to be responsive to the dipsogenic effect of angiotensin II (Felix and Akert 1974). After transection of the neural connections of the subfornical organ, Lind et al. (1983) observed a diminished pressor response to intravenously infused angiotensin II. The role of the few neurohypophyseal-peptide-containing fibers in the subfornical organ is yet unclear.

5.1.5 Area Postrema

Although the area postrema is densely surrounded by fibers and terminals containing neurohypophyseal peptides (Fig. 10 a, b) only occasionally have such fibers entering the hemoneural tissue of the area postrema been observed. Other afferent fibers to the area postrema of the rat contain substance P, adrenaline, and probably noradrenaline. Enkephalin and noradrenaline are contained in neurons of the area postrema (Armstrong et al. 1981). Axons and axon terminals containing substance P within the area postrema were identified at the ultrastructural level (D. M. Armstrong et al. 1982). Vigier and Rouvière (1979) and Vigier and Portalier (1979) described afferent connections to the area postrema from the nucleus ambiguus, efferent connections from the area postrema to the mesencephalic trigeminal nucleus, locus coeruleus, central gray, and inferior and superior colliculi, and reciprocal connections between the area postrema and the nucleus of the solitary tract, dorsal vagal nucleus, nucleus intercalatus, and hypoglossal nucleus. These fiber connections demonstrated with tracing techniques coincide with catecholaminergic fibers (Lindvall and Björklund 1974).

Van der Kooy and Koda (1983) have demonstrated, by means of anterograde and retrograde transport studies, connections between the area postrema and the internal solitary zone and dorsal division of the medial solitary nucleus, the dorsal parabracial nuclei, dorsolateral tegmental nucleus, pericentral division of the dorsal tegmental nucleus. Furthermore, a direct afferent projection from the dorsal hypothalamic area to the area postrema of the rat was traced by Hosoya and Matsushita (1981). Via these connections, information on blood levels of peptides, amines, and other compounds which can enter the area postrema, though

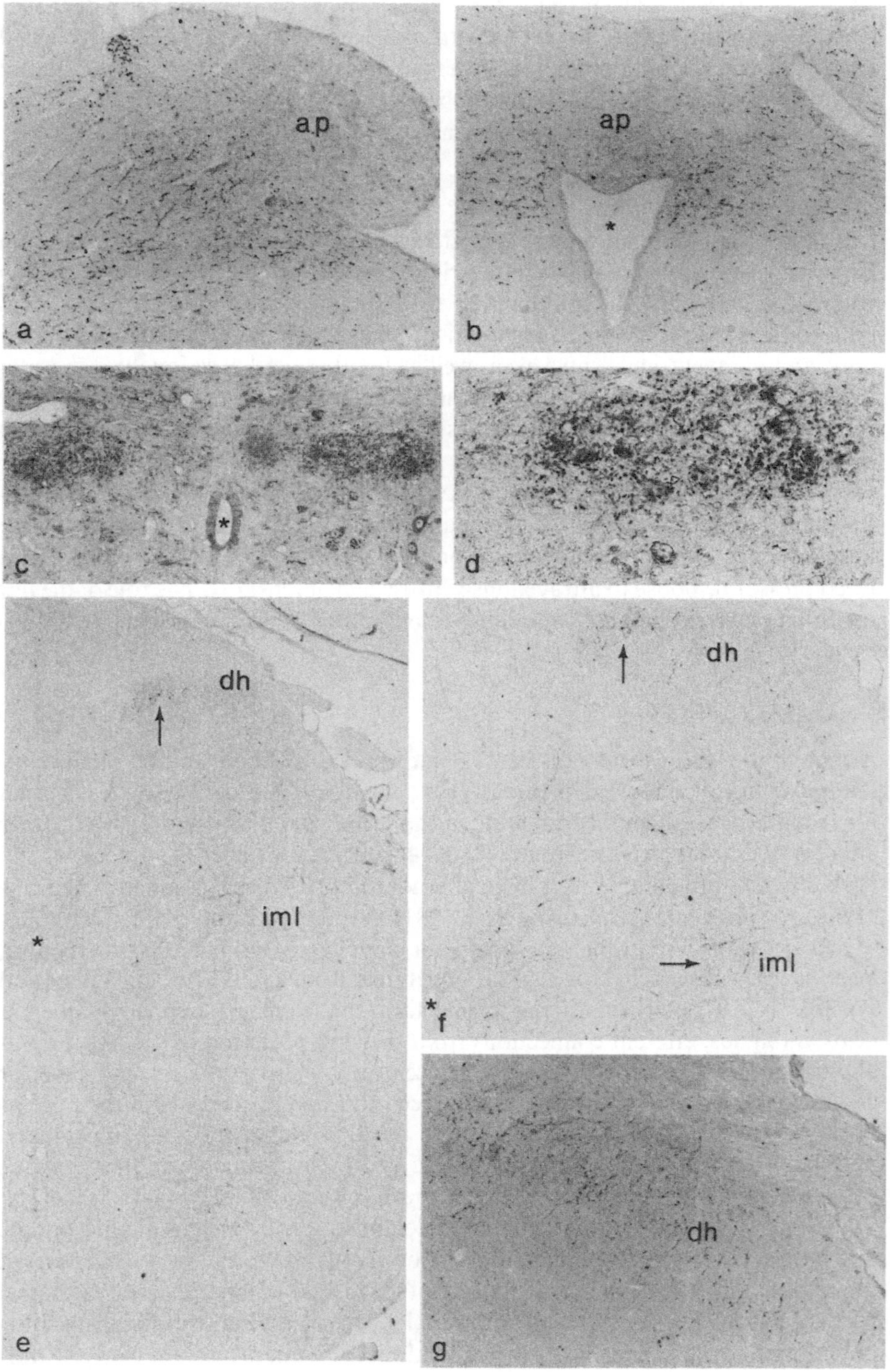
ap
a
ap
*
b
*
c
d
dh
iml
*
e
dh
iml
*
f
dh
g

not the surrounding tissue of the brain stem, may be transmitted to these autonomic regulatory centers. These centers receive, in turn, a rich input of oxytocin and vasopressin fibers from the paraventricular nucleus (Swanson 1977; Swanson and Hartman 1980; Nilaver et al. 1980; Sofroniew and Schrell 1981).

While some reports favor an important role of the area postrema in cardiovascular regulation (Michelini et al. 1983; Ferrario et al. 1979; Barnes et al. 1979), this has been disproven by Haywood et al. (1980). Excision of the area postrema did not cause hypertension (Zandberg et al. 1977).

The area postrema appears to be a site where food intake is influenced (Hyde and Miselis 1983). CCK decreases food intake. Thermal lesions of the area postrema attenuated CCK-induced reduction of food intake (Van der Kooy 1984). Ablation of the area postrema exaggerated consumption of preferred foods in the rat (Edwards and Ritter 1981). Serotonin applied topically to the area postrema of the cat induced EEG synchronization, pointing to a chemotrigger mechanism (Bronzino et al. 1972).

The area postrema is a chemoreceptor for substances of widely disparate chemical structures. Area postrema lesions abolish the conditioning of taste aversion by lithium chloride and scopolamine methylnitrate, and in addition may alter body weight maintenance and the ingestive response to palatable liquid food (Ritter et al. 1980). Area postrema lesions also abolish the antiaggressive effect of lithium chloride (McGlone et al. 1980).

5.1.6 Pineal Body

In the pineal gland of several species (guinea pig: Weindl and Sofroniew 1980; rat: Buijs and Pévet 1980; dog: Matsuura et al. 1983; tree shrew (Fig. 9c, d); pigeon: Weindl and Sofroniew 1982), vasopressin-, oxytocin-, and neurophysin-containing fibers were observed. The fibers enter the stalk of the pineal mainly through the habenular commissure and are distributed preferentially in the anterior part. In a review on the central pinealopetal projections, Korf and Møller (1984) state that vasopressin and oxytocin fibers from the paraventricular nucleus running laterodorsally toward the stria medullaris appear to enter the pineal via the habenular commissure. They form perivascular terminals in the pineal. In the hedgehog, Nürnberger and Korf (1981) found marked seasonal changes, correlating with hibernation, in the ratio of vasopressin to oxytocin fibers and in the total amount of both fiber types.

As no vasopressin, oxytocin, or neurophysin immunoreactivity is observed within pinealocytes, it is unlikely that these peptides are synthesized within the organ.

Fig. 10 a–g. *Tupaia.* Neurophysin immunoperoxidase reaction. Frontal sections through **a, b** area postrema (*ap*), **c, d** dorsal medulla oblongata, **e–g** transversal sections through spinal cord. **a, b** Some neurophysin fibers of the nucleus solitarius-vagus complex are in contact with the ap. **a–c** 80x; **d** detail of **c**, 200x; **e** survey of spinal cord, 30x; **f** detail of **a** 40x; **g** dorsal horn, 80x. Neurophysin fibers (*arrows*) are accumulated in dorsal horn (*dh*) and intermediolateral area (*iml*); * central canal

Reinharz et al. (1974) and Reinharz and Vallotton (1977) have measured neurophysin in the pineal by radioimmunoassay. Earlier reports of high levels of arginine-vasotocin in the fetal pineal organ (Pavel et al. 1973) could not be confirmed by other investigators (Negro-Vilar et al. 1979; Dogterom et al. 1980; Pévet et al. 1980). The function of the perivascular vasopressin and oxytocin fibers in the pineal has been associated with an inhibitory influence on the *N*-acetyltransferase activity (Namboodiri et al. 1981).

5.1.7 Subcommissural Organ

The subcommissural organ consists of modified secretory ependymal cells which discharge apical droplets containing neutral mucopolysaccharides into the cerebrospinal fluid. The secretory globules are transformed into filaments forming Reissner's fiber, which can be traced through the aqueduct and the fourth ventricle to the lumbosacral end of the central canal of the spinal cord.

Vasopressin, oxytocin, and neurophysin are found in close proximity to the subcommissural organ, occasionally in the hypendyma of the subcommissural organ passing toward the pineal organ (Buijs and Pévet 1980; Matsuura et al. 1983). They appear not to participate in the innervation of the subcommissural organ, which receives serotonin (Møllgard and Wiklund 1979; Bouchaud 1979) and GABA fibers (Oertel et al. 1983).

5.1.8 Choroid Plexus

Although occasionally nerve fibers containing neurohypophyseal peptides have been observed in the area of the choroid fissure, caution must be used in deducing, from these rather sporadic observations, the existence of a "hypothalamo-choroidal tract" (Brownfield and Kozlowski 1977) which may be essential in the regulation of water transport between blood and cerebrospinal fluid analogous to the epithelium of the distal tubules of the kidney (Schultz et al. 1977).

5.2. Neural Target Areas

5.2.1 Distribution of Fibers and Terminals Containing Neurohypophyseal Peptides

The distribution of neural target areas of vasopressin-, oxytocin-, and neurophysin-containing fibers and terminals is listed in Table 3. The compilation of these data is based on several literature reports (Sofroniew 1980, 1982, 1983 a, b; Sofroniew and Weindl 1981; Buijs 1983). The original observations, mostly obtained in the rat, were confirmed in many other mammalian species, including subhuman primates (Weindl et al. 1981) (Figs. 11, 12) and man (Sofroniew et al. 1981). Vasopressin and oxytocin fibers and terminals are observed in the entire neuraxis, from the olfactory bulb to the caudal end of the spinal cord, but there are variations in densities, as well as in the ratio of vasopressin and oxytocin fibers. Only scattered fibers are encountered in the cerebral cortex, and the

Table 3. Distribution of fibers and terminals containing vasopressin (VP) and oxytocin (OT) in the mammalian central nervous system

	VP	OT	Origin	References
Telencephalon				
Olfactory bulb		(+)		
Frontal cortex	(+)	(+)		
Diagonal tract of Broca	++	(+)	NIST	De Vries and Buijs (1983)
Nucleus accumbens	(+)	(+)		
Lateral septum	++++		NIST	De Vries and Buijs (1983)
Medial septum		(+)		
Central, basal, lateral, anterior, cortical amygdala	+	+		
Medial amygdala	++	+		
Nucleus interstitialis striae terminalis	+	(+)		
Dorsal hippocampus	(+)	(+)		
Ventral hippocampus	+	+		
Diencephalon				
Lateral habenula	++++		NIST	De Vries and Buijs (1983)
Mediodorsal thalamus	+++	(+)	SCN	Hoorneman and Buijs (1982)
Periventricular nucleus	++			
Dorsomedial hypothalamus	++		SCN	Hoorneman and Buijs (1982)
Posterior hypothalamus	++		SCN	Hoorneman and Buijs (1982)
Supramammillary nucleus	(+)	+		
Mesencephalon				
Substantia grisea	+	+	NIST	De Vries and Buijs (1983)
Nucleus raphes dorsalis	+	+		
Substantia nigra pars compacta	++	++++		
Ventral tegmental area	+	++		
Nucleus interpeduncularis	(+)	+		
Rhombencephalon				
Periventricular gray	+	+		
Nucleus parabrachialis dorsalis	+	++		
Nucleus parabrachialis ventralis	(+)	+		
Locus coeruleus	+	+	NIST	De Vries and Buijs (1983)
Nucleus raphes pontis	(+)	(+)		
Nucleus tractus solitarii	++	++++	PVN	Swanson and Kuypers (1980)
Nucleus dorsalis vagi	++	++++	PVN	Swanson and Kuypers (1980)
Nucleus commissuralis	++	++++	PVN	Swanson and Kuypers (1980)
Nucleus ambiguus	+			
Nucleus reticularis lateralis	+	++		
Nucleus raphes magnus	+	++		
Nucleus raphes obscurus	(+)	(+)		
Substantia gelatinosa nervi trigemini	(+)	+		
Spinal cord				
Dorsal horn (lamina I–III)	(+)	+	PVN	Swanson and Kuypers (1980)
Central gray (lamina X)	+	++		
Nucleus intermediolateralis	(+)	+	PVN	Swanson and Kuypers (1980)

Abbreviations: NIST, nucleus interstitialis striae terminalis; PVN, paraventricular nucleus; SCN, suprachiasmatic nucleus

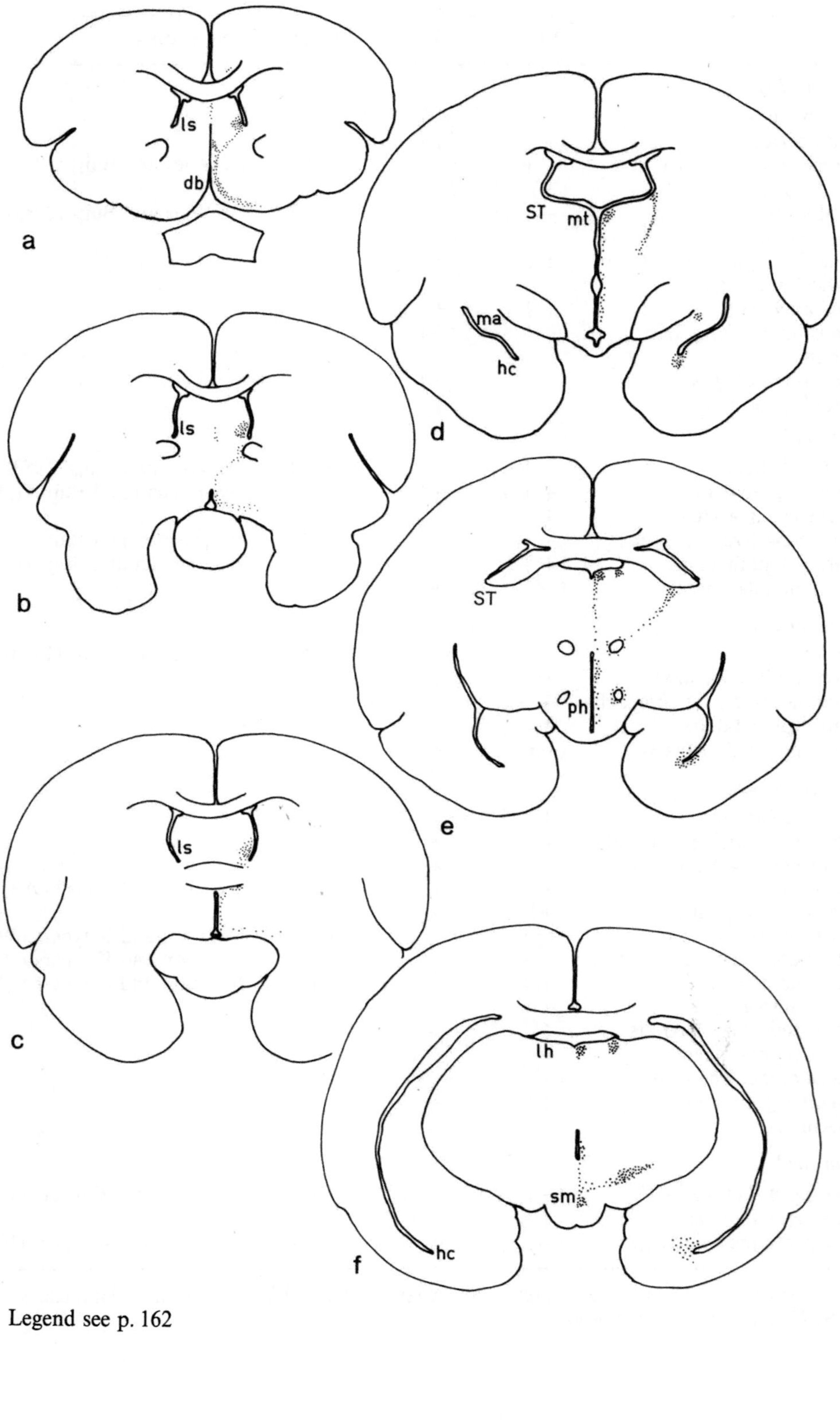

Legend see p. 162

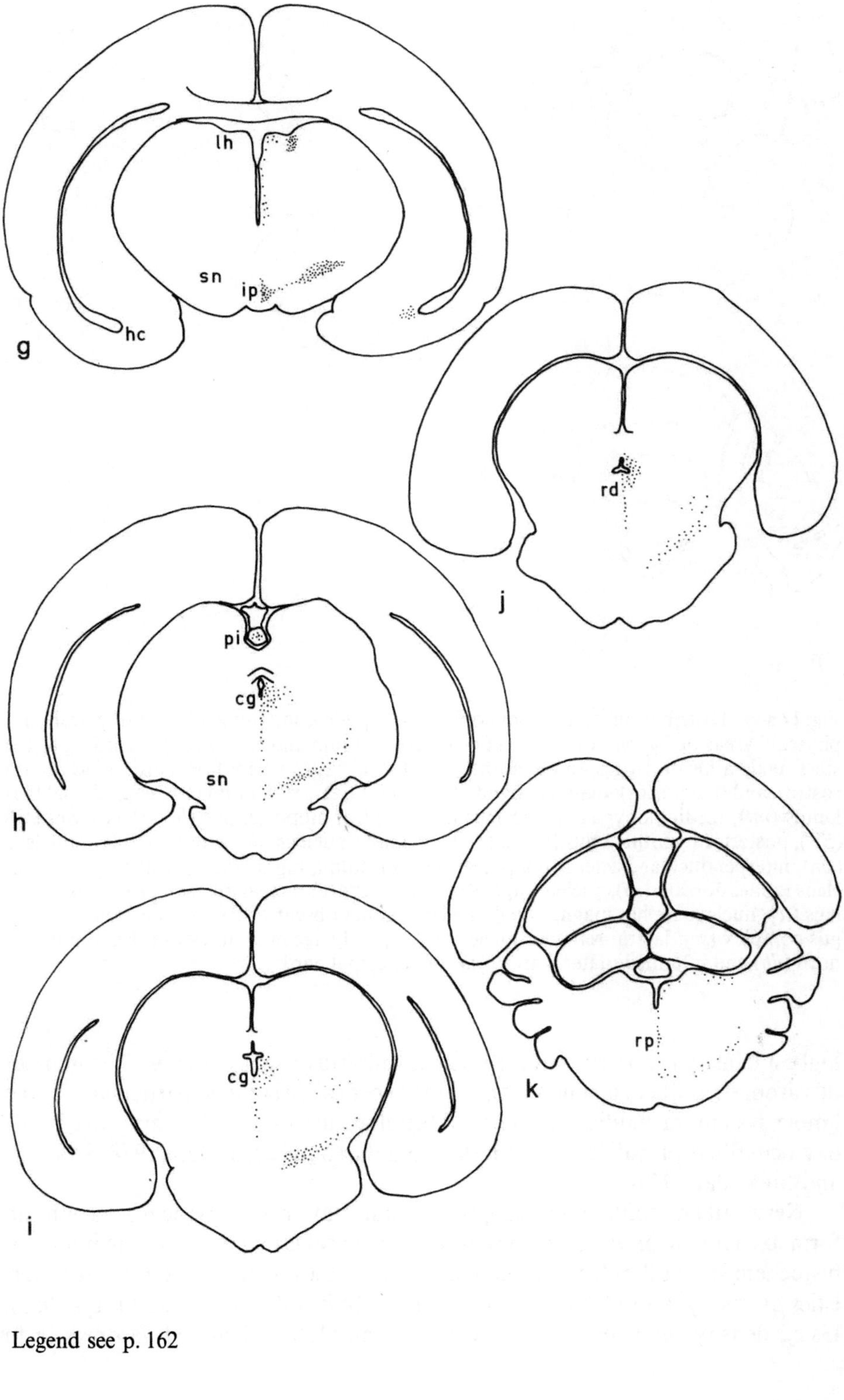

Legend see p. 162

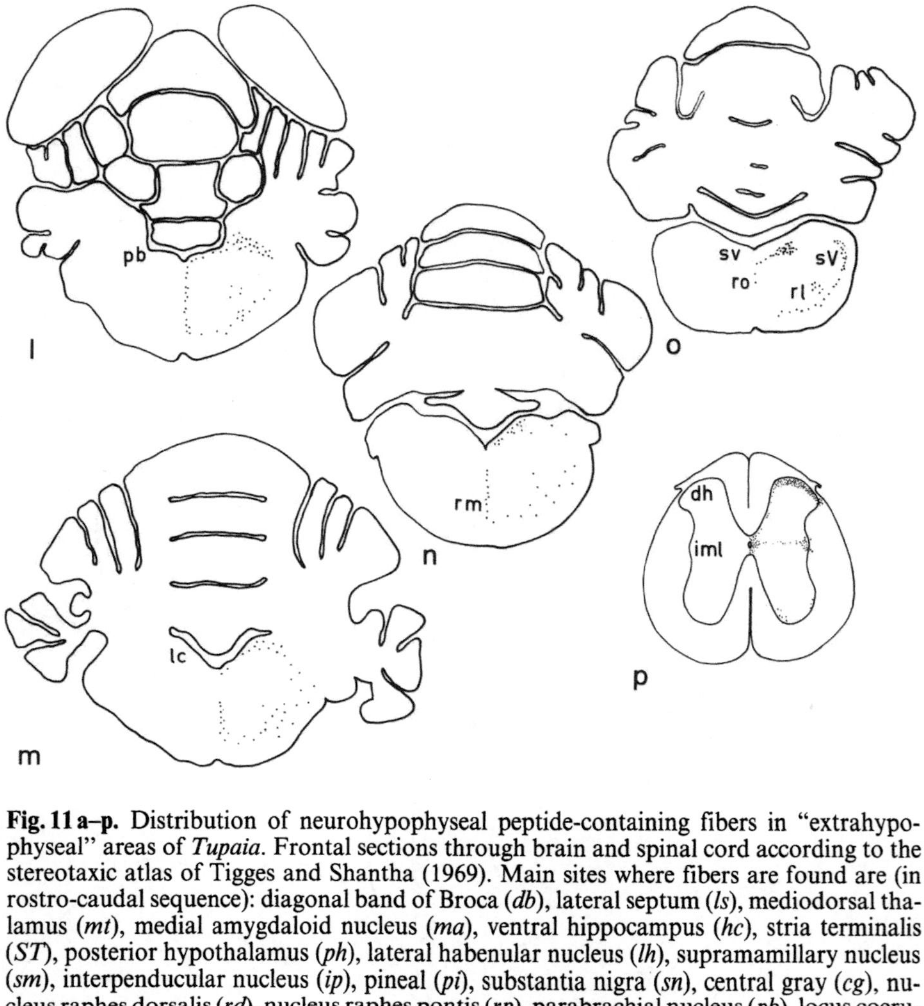

Fig. 11 a–p. Distribution of neurohypophyseal peptide-containing fibers in "extrahypophyseal" areas of *Tupaia*. Frontal sections through brain and spinal cord according to the stereotaxic atlas of Tigges and Shantha (1969). Main sites where fibers are found are (in rostro-caudal sequence): diagonal band of Broca (*db*), lateral septum (*ls*), mediodorsal thalamus (*mt*), medial amygdaloid nucleus (*ma*), ventral hippocampus (*hc*), stria terminalis (*ST*), posterior hypothalamus (*ph*), lateral habenular nucleus (*lh*), supramamillary nucleus (*sm*), interpenducular nucleus (*ip*), pineal (*pi*), substantia nigra (*sn*), central gray (*cg*), nucleus raphes dorsalis (*rd*), nucleus raphes pontis (*rp*), parabrachial nucleus (*pb*), locus coeruleus (*lc*), nucleus raphes magnus (*rm*), nucleus raphes obscurus (*ro*), nucleus solitarius-vagus complex (*sv*), lateral reticular nucleus (*rl*), spinal trigeminal nucleus (*sV*), and dorsal horn (*dh*) and intermediolateral area (*iml*) of the spinal cord

highest density is observed in the nucleus solitarius/vagus complex. The number of vasopressin fibers is much higher in subcortical structures, particularly in the limbic system, including the lateral habenular nucleus and the amygdala, while oxytocin fibers prevail in the brain stem and the spinal cord (Buijs 1978; Swanson and McKellar 1979).

Nerve fibers containing vasopressin/neurophysin or oxytocin/neurophysin form axosomatic or axodendritic contacts in neural target areas. Using immunohistochemistry at the electron-microscopic level, these contacts exhibit characteristics of true synapses (Buijs and Swaab 1979; Buijs 1983; Voorn and Buijs 1983). Using density centrifugation, Hawthorn et al. (1984) identified vasopressin in

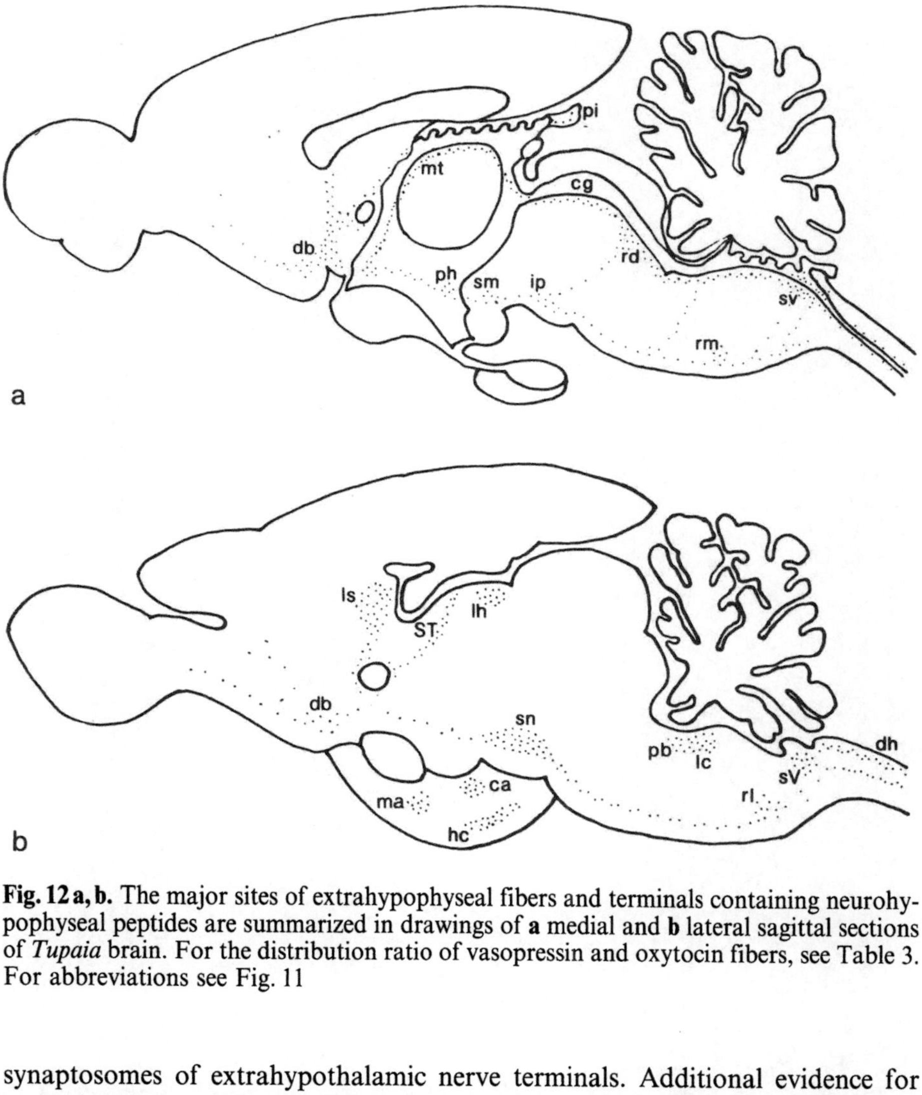

Fig. 12a, b. The major sites of extrahypophyseal fibers and terminals containing neurohypophyseal peptides are summarized in drawings of **a** medial and **b** lateral sagittal sections of *Tupaia* brain. For the distribution ratio of vasopressin and oxytocin fibers, see Table 3. For abbreviations see Fig. 11

synaptosomes of extrahypothalamic nerve terminals. Additional evidence for synaptic release of vasopressin and oxytocin in extrahypothalamic target areas was obtained by Buijs and van Heerikhuize (1982).

Within the *telencephalon,* the olfactory bulb contains some oxytocin fibers. The frontal, piriform, cingulate, and entorhinal cortices contain only scattered vasopressin and oxytocin fibers. Few vasopressin and oxytocin fibers are found in the nucleus accumbens, slightly more in the ventral hippocampus (Fig. 13) and in the central, anterior, basal, cortical and lateral amygdala. In the medial amygdala and bed nucleus of the stria terminalis, more vasopressin than oxytocin fibers are encountered. The nucleus of the diagonal band of Broca (Fig. 9a) and the lateral septum (Fig. 14d–f) contain considerably higher amounts of vasopressin than oxytocin fibers and terminals.

In the *diencephalon,* high numbers of vasopressin and few oxytocin terminals are contained in the mediodorsal thalamus. The lateral habenular nuclei (Fig. 15)

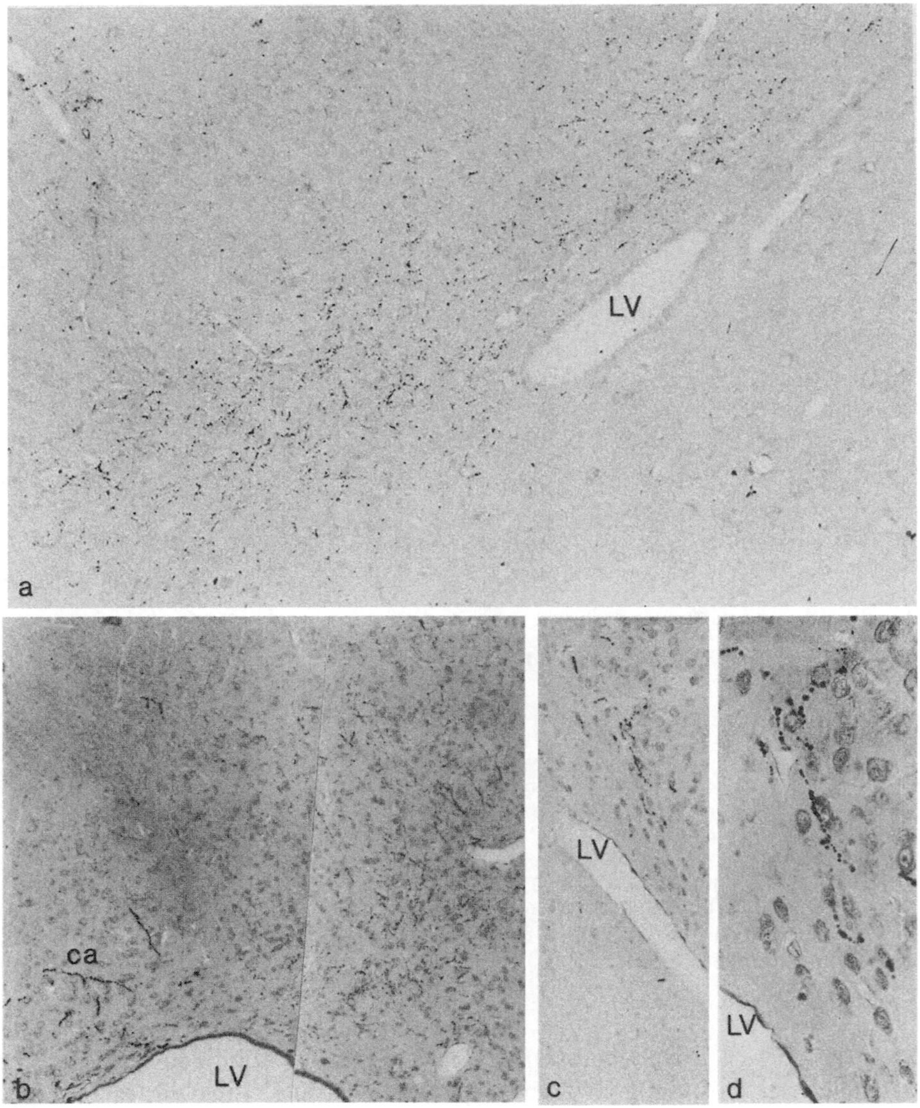

Fig. 13. a, b *Tupaia;* **c, d** *Sanguinus fuscicollis.* Neurophysin immunoperoxidase reaction (toluidin blue). **a, c, d** Frontal and **b** horizontal sections through temporal lobe. Small caliber neurophysin fibers are present in the ventral **a–d** hippocampus, large caliber fibers in the **b** central amygdaloid nucleus (*ca*). *LV,* lateral ventricle. **a** 160x; **b, c** 80x; **d** detail of **c** (200x)

and the posterior and periventricular nuclei contain high numbers of vasopressin but no oxytocin fibers. Few vasopressin and some oxytocin fibers are found in the supramammillary nucleus.

At the level of the *mesencephalon,* vasopressin and oxytocin fibers are present in the central gray area and the dorsal raphe nucleus. More oxytocin than vaso-

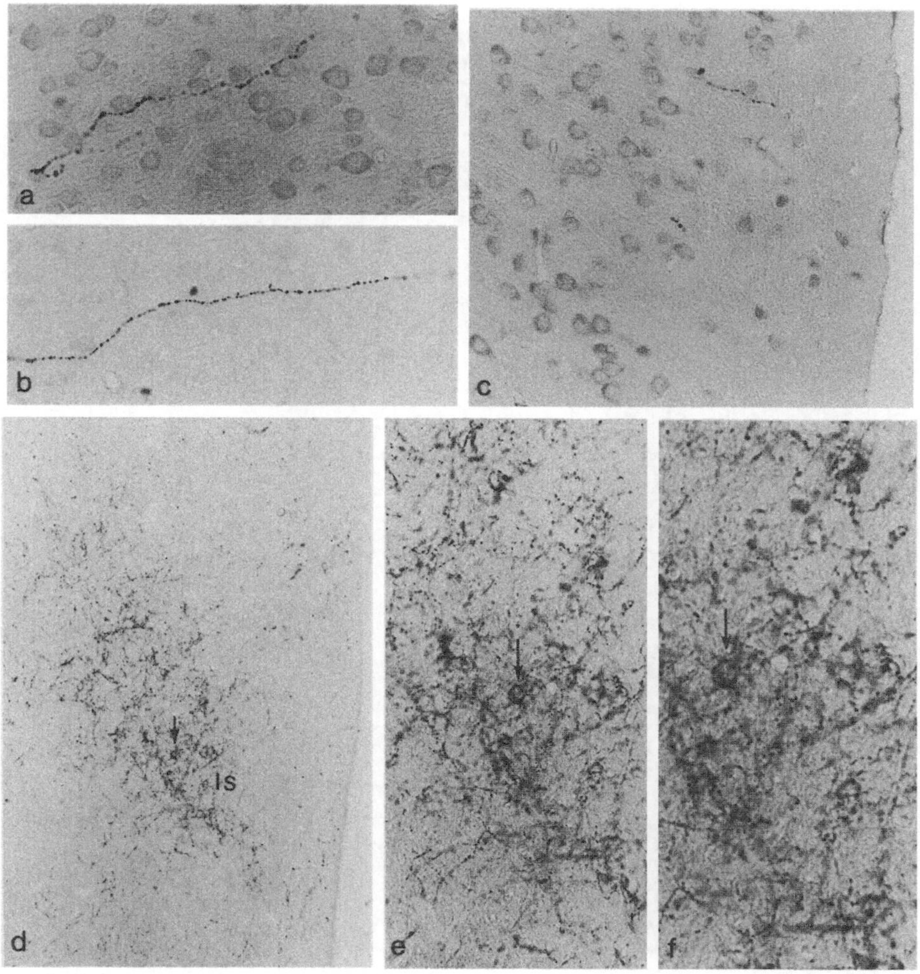

Fig. 14, a–c *Sanguinus fuscicollis*. **d–f** *Tupaia*. Neurophysin immunoperoxidase reaction (toluidin blue). **a–c** Frontal sections through rostral telencephalon. Neurophysin fibers are present at several rostrocaudal levels (200x). **d–f** Frontal section through lateral septum (*ls*). **d** Survey (80x); **e, f** details of **d** (**e** 200x; **f** 320x). Neurophysin fibers terminate in ls. ↓ axo-somatic contact

pressin fibers and terminals are found in the substantia nigra pars compacta, ventral tegmental area, and interpeduncular nucleus.

In the *rhombencephalon,* more oxytocin than vasopressin fibers are found in the parabrachial nuclei, nucleus raphes magnus, nucleus solitarius/vagus complex, nucleus commissuralis, nucleus reticularis lateralis, and substantia gelatinosa nervo trigemini. The locus coeruleus and nucleus raphes pontis contain few vasopressin and oxytocin fibers.

In the *spinal cord* (Fig. 10e–g), more oxytocin than vasopressin fibers are found in the dorsal horn, central gray (lamina X), and in the intermediolateral nucleus.

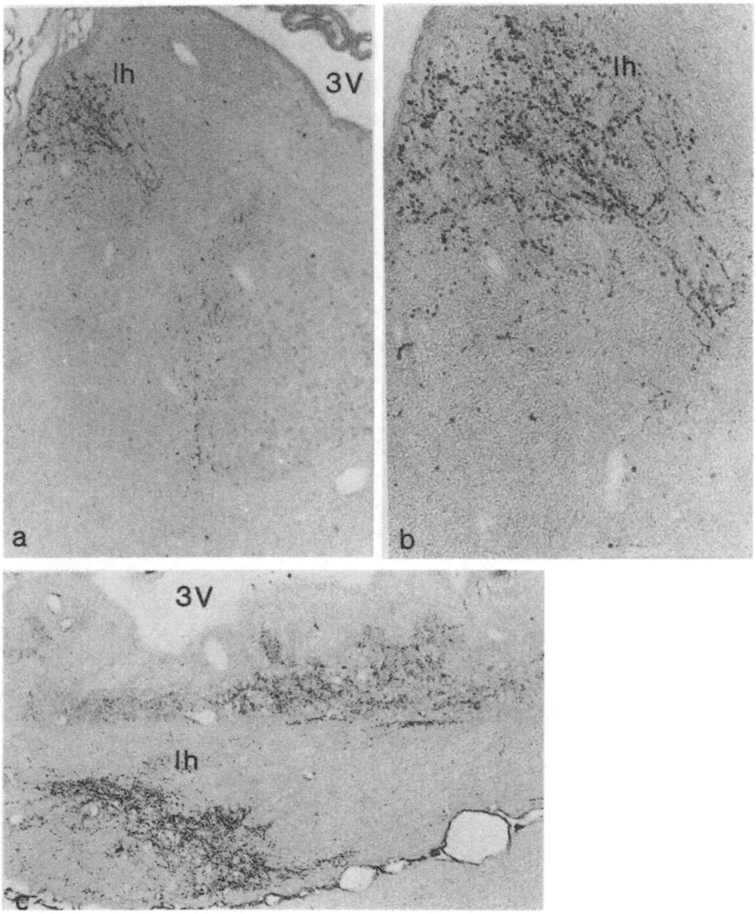

Fig. 15 a–c. *Tupaia*. **a, b** Frontal and **c** horizontal sections through lateral habenular nucleus (*lh*). Neurophysin immunoperoxidase reaction showing dense terminal fields in lh. 3V, 3rd ventricle. **a** 80x; **b** 200x; **c** 120x

This specific pattern of a very wide distribution of vasopressin and oxytocin fibers and terminals appears well-established by observations from several independent laboratories.

Various earlier reports on the origin and the termination of projections containing neurohypophyseal peptides to neural target areas were based mainly on observations of consecutive or closely neighboring immunostained sections (Buijs 1978; Buijs et al. 1978; Sofroniew and Weindl 1978; Sofroniew 1980; Yulis and Rodriguez 1982). On the basis of subsequent tract-tracing studies combined with immunohistochemical identification of neurohypophyseal peptides on alternating sections, lesion studies, and observations of additional parvocellular vasopressin/neurophysin-containing perikarya which could be detected only after colchicine treatment (Van Leeuwen and Caffé 1983; Caffé and Van Leeuwen 1983), not all of the originally proposed pathways could be confirmed.

5.2.2 Fiber Projections Containing Neurohypophyseal Peptides

In an extensive review of the organization of the paraventricular and supraoptic nuclei and their connections, Swanson and Sawchenko (1983) report that neurons of the supraoptic nucleus project only to the neural lobe. However, Alonso et al. (1984) reported that anterograde radioautographic tracing revealed fiber projections from magnocellular vasopressin/neurophysin perikarya rostrally to the olfactory bulb, cingular and cerebral cortices, medial nucleus of the amygdala (through the stria terminalis), lateral habenular nucleus, subcommissural organ (through the stria medullaris), and mammillary bodies.

The main source of the extrahypothalamic fiber projections appears to be the paraventricular nucleus (Swanson and Kuypers 1980; Swanson and Sawchenko 1983). Earlier investigations had demonstrated descending pathways from the paraventricular nucleus to the spinal cord (Kuypers and Maisky 1975; Saper et al. 1976; Hancock 1976; Cratcher et al. 1978; Kneisley et al. 1978; Ono et al. 1978; Blessing and Chalmers 1979; Hosoya and Matsushita 1979) and to the dorsal vagal nucleus and nucleus of the solitary tract (Saper et al. 1976; Conrad and Pfaff 1976). Swanson and Kuypers (1980) observed that most neurons projecting to the neural lobe are confined to the parvocellular division and that most neurons projecting to the neural lobe are confined to the magnocellular division. Some neurons may project to both target areas; this possibility is favored by electrophysiological data (Zerihun and Harris 1981).

The neurons of the paraventricular nucleus projecting to the spinal cord are concentrated in the dorsal, lateral, and medial (ventral half) parts of the parvocellular division of the nucleus and in the posterior part of the magnocellular division (Swanson and Kuypers 1980). Projections from the dorsal vagal complex derive from similarly distributed cells, except for a very few cells in the dorsal parvocellular group which contribute to this pathway. Some individual cells in the ventral half of the medial parvocellular division and in the lateral parvocellular part of the paraventricular nucleus innervate both the sympathetic and parasympathetic cell groups.

Immunohistochemical studies by Swanson (1977) with differentially cross-absorbed antisera to neurophysin I and II in the cow and rat demonstrated pathways to the dorsal vagal complex and spinal cord. The fibers descend through the median forebrain bundle, pass between the substantia nigra and the red nucleus, continue through the ventrolateral parts of the reticular formation, and enter the dorsolateral funiculus of the spinal cord. In the pons, fibers branch off to the parabrachial nucleus and locus coeruleus. In the medulla oblongata, fibers bend dorsally towards the dorsal vagal nucleus and the nucleus of the solitary tract. A second pathway descending through the central gray appears to innervate the Edinger-Westphal nucleus, locus coeruleus, and central gray. Immunohistochemical studies by Sofroniew and Weindl (1978), Buijs (1978), Swanson and Hartman (1980) indicated that considerably more oxytocin than vasopressin fibers innervate both brain stem and spinal cord. Swanson and Sawchenko (1983), in contrast to Nilaver et al. (1980) found that oxytocin fibers predominate in the dorsal vagal complex and spinal cord. Oxytocin terminals are concentrated in the medial division of the nucleus of the solitary tract at the level of the area postrema, and are concentrated in the medial and lateral tips of the dorsal vagal nucleus. Buijs (1983) demonstrated oxytocin within round clear vesicles of asymmetrical synapses in the nucleus of the solitary tract.

A detailed account of the projections to the spinal cord was given by Swanson and McKellar (1979) in the rhesus monkey and rat. Oxytocin fibers in the spinal cord course in the dorsolateral funiculus to the filum terminale. Within the gray matter they terminate at neurons in the intermediolateral column, the central gray, and the marginal zone of the dorsal horn. These locations suggest that fibers in the intermediolateral column are possibly involved in autonomic regulation and that fibers in the substantia gelatinosa of the dorsal horn and the spinal trigeminal nucleus are possibly involved in processing nociceptive and thermic information.

Direct evidence for these fibers projections was obtained from combined retrograde tracing and immunohistochemical studies (Sawchenko and Swanson 1982a, b; Sofroniew and Schrell 1981). After injection of tracer into the dorsal vagal complex or upper thoracic segments of the spinal cord, all labeled neurons were confined to the paraventricular nucleus, and 3–4 times more oxytocin than vasopressin neurons were retrogradely labeled. Only about 20% of the retrogradely labeled neurons in the paraventricular nucleus contained vasopressin and oxytocin. Other descending projections from the paraventricular nucleus derive from neurons containing tyrosine hydroxylase (dopaminergic neurons) (Swanson et al. 1981), somatostatin, metenkephalin and from yet unidentified neurons (Swanson and Sawchenko 1983).

Results of combined tract-tracing studies and immunohistochemical studies from several laboratories (Sawchenko and Swanson 1982a; Armstrong et al. 1980; Sofroniew and Schrell 1981) indicate that from a more caudally located subgroup of the paraventricular nucleus, vasopressin/neurophysin- and oxytocin/neurophysin-containing fibers project to the mesencephalic gray, dorsal raphe nucleus, nucleus of the solitary tract, dorsal vagal nucleus, commissural nucleus ("dorsal vagal complex"), lateral reticular nucleus, and dorsal column ("substantia gelatinosa", zone II–III of Rexed, 1952) and intermediolateral column of the spinal cord. Vasopressin innervation is limited to the nucleus solitarius/vagus complex and the substantia gelatinosa. The solitary tract nucleus receives very rich innervation from peptidergic neurons. In addition to neurohypophyseal peptides, somatostatin, substance P, and enkephalin terminals exhibit distinct patterns of distribution in the various subnuclei (Kalia et al. 1984).

The number of target areas of vasopressin/neurophysin projections from the suprachiasmatic nucleus was found to be much smaller than postulated in earlier reports (Buijs 1978; Sofroniew and Weindl 1978) when additional techniques such as lesions of the suprachiasmatic nuclei (Hoorneman and Buijs 1982) or retrograde transport of horseradish peroxidase (Schrell et al. 1983) were used.

From the parvocellular vasopressin/neurophysin perikarya of the suprachiasmatic nucleus, fibers project to the medial dorsal thalamus (nucleus periventricularis rotundocellularis and stellatocellularis thalami) and the nucleus ventromedialis and dorsomedialis hypothalami (Schrell et al. 1983).

A projection of vasopressin/neurophysin fibers from the suprachiasmatic nucleus to the OVLT in the rat was reported by Buijs (1978). According to the observations by Weindl and Sofroniew (1981), vasopressin/neurophysin fibers in the OVLT of the rat appear to run through the internal zone and to make only few contacts with blood vessels of the external zone. No vasopressin/neurophysin fibers project from the suprachiasmatic nucleus to the external zone of the median eminence (Vandesande et al. 1977).

Additional parvocellular vasopressin/neurophysin perikarya observed after colchicine treatment and not present in homozygous Brattleboro rats (Caffé and Van Leeuwen 1983; Van Leeuwen and Caffé 1983) appear to be a source of additional extrahypothalamic vasopressin/neurophysin fiber projections.

According to lesion and dye injection studies of De Vries and Buijs (1983), vasopressin/neurophysin fibers from the bed nucleus of the stria terminalis project to the lateral septum, diagonal band of Broca, anterior amygdala, lateral habenular nucleus, periventricular gray and locus coeruleus. De Vries et al. (1981) reported much higher densities of vasopressin/neurophysin terminals in the lateral septum of male as compared to female rats, a difference becoming evident from postnatal day 10 onwards.

Kawata et al. (1983 b) reported that thick fibers project to vascular and thin fibers to neural target areas.

In a recent survey, Buijs (1983) has summarized the major vasopressin and oxytocin pathways in the rat brain (Fig. 16 a, b). Vasopressin/neurophysin neurons of the paraventricular nucleus project in a dorsal pathway to the locus coeruleus, dorsal raphe nucleus, dorsal vagal complex, and spinal cord, and in a ventral pathway to the locus coeruleus and nucleus ambiguus. However, the origin of vasopressin/neurophysin fibers projecting to amygdala, ventral hippocampus and ventral tegmental area is uncertain. Oxytocin/neurophysin fibers in the olfactory bulb appear to arise from magnocellular perikarya of the paraventricular nucleus, and the major origin of these fibers altogether are these perikarya. Oxytocin/neurophysin fibers from them appear to project to the medial septum, amygdala and superior colliculi. In a dorsal pathway, fibers project to the periventricular gray, parabrachial nuclei, dorsal vagal complex, and spinal cord, and in a ventral pathway (running in the median forebrain bundle) fibers project to the substantia nigra, nucleus ambiguus and spinal cord. The same pattern of extrahypothalamic oxytocin/neurophysin fiber distribution was observed in homozygous diabetes insipidus rats (Nilaver et al. 1982). Branches of oxytocin/neurophysin fibers entering the cerebellum were reported only by Buijs (1978).

Not for all of these connections reported in the survey by Buijs (1983) has final evidence been established. Sufficient certainty exists, however, for the following fiber projections: oxytocin and vasopressin fibers from the paraventricular nucleus to the dorsal raphe nucleus, mesencephalic gray, nucleus solitarius/vagus complex, lateral reticular nucleus, substantia gelatinosa and intermediolateral column of the spinal cord; vasopressin fibers from the parvocellular suprachiasmatic nucleus to the mediodorsal thalamus and posterior hypothalamus, as well as from parvocellular vasopressin neurons of the bed nucleus of the stria terminalis to the lateral septum.

Electrophysiological studies support fiber connections from neurosecretory nuclei to extrahypothalamic targets in the brain (Pittman et al. 1981) and to the spinal cord (Yamashita et al. 1984).

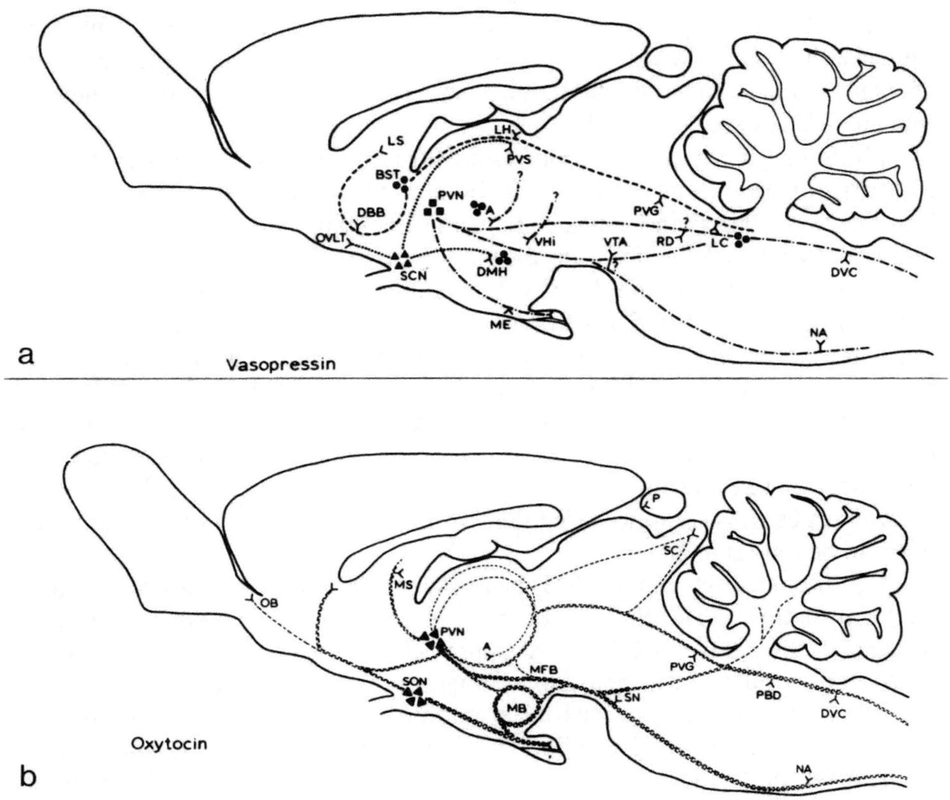

Fig. 16 a, b. Vasopressin and oxytocin fiber projections in the rat brain. **a** The major vasopressinergic pathways in the brain with their most likely origin (see text). The pathways of the paraventricular nucleus (*PVN*) are indicated by (—·—). Pathways of the suprachiasmatic nucleus (*SCN; triangles*) are indicated by (.....). The vasopressinergic cell groups found only after colchicine treatment are indicated by *black dots,* and the pathways of the bed nucleus of the stria terminalis (*BST*) are indicated by (– – –). The *question marks* indicate areas where the source of the vasopressin innervation is still unknown. **b** The major oxytocinergic pathways largely arising from the PVN (see text). *A,* amygdala; *DBB,* diagonal band of Broca; *DMH,* dorsomedial nucleus of the hypothalamus; *DVC,* dorsal vagal complex; *LC,* locus coeruleus; *LH,* lateral habenula; *LS,* lateral septum; *MB,* mammillary body; *ME,* median eminence; *MFB,* medial forebrain bundle; *MS,* medial septum; *NA,* nucleus ambiguus; *OB,* olfactory bulb; *OVLT,* organum vasculosum laminae terminalis; *P,* pineal; *PBD,* dorsal parabrachial nucleus; *PVG,* periventricular gray; *PVS,* periventricular nucleus; *RD,* dorsal raphe nucleus; *SC,* superior colliculus; *SN,* substantia nigra; *SON,* supraoptic nucleus; *VHi,* ventral hippocampus; *VTA,* ventral tegmental area. From Buijs (1983)

5.2.3 Ontogenetic Aspects

Vasopressin appears in magnocellular perikarya of the *rat* around fetal day 16 (Choy and Watkins 1979; Watkins and Choy 1979). Buijs et al. (1980 a, b) found vasopressin-containing cells from fetal day 16 in the supraoptic nucleus and from day 18 in the paraventricular nucleus. Similar observations were made by Wolf and Trautmann (1977), using neurophysin staining. At fetal day 17, fibers grow out to the median eminence and olfactory bulb. The immunohistochemical data are largely confirmed by the author's radioimmunological measurements. Gash et al. (1980), using immunohistochemistry, electron microscopy, and vasopressin and oxytocin radioimmunoassay, observed a similar time schedule. The maturation of neurophysin perikarya in the supraoptic nucleus is accompanied by catecholamine innervation (Sladek et al. 1980). The postnatal development was studied by Burlet et al. (1979) and Krisch (1980).

Vasopressin perikarya in the suprachiasmatic nucleus were observed on postnatal day 2 by De Vries et al. (1981). Using combined immunohistochemistry and ^{3}H-thymidine autoradiography, Okamura et al. (1983) found vasopressin perikarya in the supraoptic nucleus between days 12 and 14 of gestation.

A similar time sequence is found in the *mouse,* where neurophysin neurons are observed first in the supraoptic nucleus at fetal day 13½, and neurophysin synapses between postnatal days 5 and 15 (Silverman et al. 1980b).

In the hypothalamoneurohypophyseal system of the *pig,* neurophysin neurons develop at day 60 (Ellis and Watkins 1975). Parvocellular vasopressin/neurophysin perikarya in the suprachiasmatic nucleus were demonstrated at day 80, vasopressin/neurophysin fibers in the external zone of the median eminence at day 100 (Weindl et al. 1984).

In the fetal *human* brain, the first oxytocin-neurophysin neurons were seen as early as the 14th week of gestation (Paulin et al. 1978). Radioimmunoassayable oxytocin was detected in fetal pituitary glands in the 14th week of gestation (Khan-Dawood and Dawood 1984) and vasopressin in pituitary and hypothalamus in the 15th week of gestation (Schubert et al. 1981). Impaired postnatal brain development, especially of cerebellum and medulla oblongata, was observed in diabetes insipidus rats (Boer et al. 1982).

In *aged rats,* Watkins and Choy (1980) observed degenerative processes in neurons of the hypothalamoneurohypophyseal system preferentially affecting oxytocin-producing neurons in the rostral paraventricular nucleus. Using radioimmunoassay, Dorsa and Bottemiller (1982) found decreased vasopressin contents in extrahypothalamic regions of senescent rats.

5.2.4 Phylogenetic Aspects

In the *lamprey* (Goossens et al. 1977a) and in several *fish* species vasotocin and isotocin were demonstrated in the hypothalamus (Goossens et al. 1977b, 1978; Schreibman and Halpern 1980) as well as in extrahypothalamic areas (Van den Dungen et al. 1982). Van den Dungen et al. (1982) demonstrated an extensive extrahypothalamic system of vasotocin and isotocin fibers.

Dierickx and Vandesande (1977) and van Vossel-Daeninck et al. (1979) demonstrated vasotocinergic and mesotocinergic neurons in the *amphibian* brain.

In the *reptilian brain,* vasotocin and mesotocin neurons were demonstrated immunohistochemically by Bons (1983) and Goossens et al. (1979).

In *birds,* neurohypophyseal peptides were demonstrated in magnocellular perikarya of the diencephalon (Berk et al. 1982; Bons 1980a, b; McNeill et al. 1976), as well as in fibers projecting to many extrahypothalamic neural target areas (Weindl et al. 1980). Noteworthy is a very extensive network of fibers forming two bands of terminals in the optic tectum of the pigeon which exhibits a cortex-like structure (Weindl and Sofroniew 1980).

5.3 Confirmation of Vasopressin, Oxytocin and Neurophysin Projections by Other Techniques

5.3.1 Quantitative Measurements of Neurohypophyseal Peptides in Discrete Areas of Brain and Spinal Cord

Quantitative measurements of vasopressin, oxytocin, and neurophysin in discrete microdissected areas of brain and spinal cord of several species have been carried out to support the immunohistochemical findings. George and Jacobowitz (1976) found vasopressin in discrete areas of the rat hypothalamus, corresponding to the hypothalamoneurohypophyseal system. Dogterom et al. (1978 a) have measured vasopressin and oxytocin by radioimmunoassay and have found high concentrations in the following major areas: septum, amygdala, hippocampus, medial-dorsal thalamus, substantia nigra, locus coeruleus, nucleus solitarius/vagus complex, substantia gelatinosa of trigeminal nerve and spinal cord, and intermediolateral nucleus of spinal cord.

These findings in the rat were confirmed by Hawthorn et al. (1980) and Glick and Brownstein (1980). A similar pattern of distribution was observed in the human central nervous system by radioimmunoassay measurements (Rossor et al. 1981; Jenkins et al. 1984). No significant changes of vasopressin could be observed in patients with Parkinson's and Huntington's disease (Rossor et al. 1982).

Although the radioimmunoassay data are very important for confirmation of the distribution pattern derived from immunohistochemical data, this method does not provide the same high degree of resolution as revealed by immunohistochemical techniques with respect to neural connectivity.

Quantitative determinations of vasopressin in the *cerebrospinal fluid* measured by bioassay as "antidiuretic activity" date back to Heller et al. (1968). Vasopressin has been determined by radioimmunoassay in the cerebrospinal fluid by Dogterom et al. (1977, 1978 b), Jenkins et al. (1980), Mens (1983), Vorherr et al. (1968), Luerssen and Robertson (1980) and by Robinson and Jones (1982).

Sodium applied to the third ventricle of the rat elicited increased vasopressin secretion into the ventricular cerebrospinal fluid (Barnard and Morris 1982). Vasopressin exhibits a rhythmic daily pattern in cerebrospinal fluid, depending on the light-dark cycle (Mens et al. 1982), but not in blood (Reppert et al. 1981). Similar patterns were observed for oxytocin and estrogen-stimulated neurophysin (Perlow et al. 1982).

Measurements of vasopressin in the cerebrospinal fluid of patients with degenerative neurological disorders and other neurological diseases did not reveal a clear-cut pattern (Sundquist et al. 1983). Measurable amounts of vasopressin, oxytocin, and neurophysin in the cerebrospinal fluid as determined by radioimmunoassay suggest that these peptides most likely enter the cerebrospinal fluid compartment from neurons and/or from fibers and terminal fields located in the vicinity of the ventricular system.

5.3.2 Receptor-Binding Studies of Neurohypophyseal Peptides in Neural Target Areas

Further support of the distribution of extrahypothalamic fiber projections containing neurohypophyseal peptides may be derived from receptor-binding studies. Specific vasopressin receptors were demonstrated in the pituitary gland by Lutz-Bucher and Koch (1983). Pearlmutter et al. (1983) showed specific binding sites for tritiated vasopressin in the rat brain, as well as regional differences. Using autoradiography, Dorsa et al. (1983) found binding of ^{3}H-vasopressin in the medulla of the kidney and in various regions of the brain, with a high density of binding sites in the nucleus of the solitary tract. Similar observations were made by Yamamura et al. (1983) and Van Leeuwen and Wolters (1983). Intense autoradiographic labeling was obtained in the lateral septum, moderate labeling in the neural lobe and the anterior lobe (Van Leeuwen and Wolters 1983).

6 Neurohypophyseal Peptides in Tissues Other Than the Central Nervous System

An increasing number of reports indicate that neurohypophyseal peptides are present in various tissues other than brain and spinal cord.

In the *retina* of rat, human, and bovine, vasopressin, oxytocin and their related neurophysins were determined by Gauquelin et al. (1983); the authors reported light-dependent changes of these peptides.

In *peripheral nerves,* including the vagal and ischiadic nerve, vasopressin immunoreactivity was demonstrated using radioimmunoassay; however, no immunohistochemical data are provided by the authors (Sequeira and Chaiken 1984).

Vasopressin immunoreactivity has also been reported in *sympathetic ganglia* of the rat (Hanley et al. 1984). Vasopressin and oxytocin elicited changes in electrical activity of neurons in the rabbit superior cervical ganglion (Wali 1984).

Oxytocin and neurophysin immunoreactive fibers were demonstrated in the *intermediate lobe of the pituitary* of the rabbit and hare at the light- and electron-microscopic level (Schimchowitsch et al. 1983a). In neither species were dopamine fibers found in the pars intermedia.

Several laboratories have demonstrated immunoreactive oxytocin and neurophysin in luteal cells of the *ovary* of various species by radioimmunoassay (Khan-Dawood et al. 1984; Wathes and Swann 1982; Flint and Sheldrick 1982; Wathes et al. 1984). Watkins (1983) and Guldenaar et al. (1984) demonstrated oxytocin and neurophysin immunohistochemically in large luteal cells of the rat ovary. Furthermore, vasopressin was demonstrated immunohistochemically in the ovary by Wathes et al. (1983). In the ovaries of homozygous diabetes insipidus rats, vasopressin was present in measurable amounts (Lim et al. 1984). This intriguing finding has been interpreted (Bonner and Brownstein 1984) by postulating that the genetic defect in producing a functionally nondeficient vasopressin precursor molecule is expressed only in neurons of the central nervous system, of the diabetes insipidus rat, not in other organs such as ovary, testis and adrenal medulla.

Other organs where neurohypophyseal peptides have been demonstrated are the testis, the adrenals, and the placenta.

In the *testis* of the rat, Pickering (1984) reported oxytocin immunoreactivity in interstitial cells (Leydig cells), and Nicholson et al. (1984) found vasopressin immunoreactivity. Kasson et al. (1984) reported high levels of radioimmunoassayable vasopressin in the rat testis.

In chromaffin cells of the *adrenal medulla,* Nussey et al. (1984) found evidence for the biosynthesis of vasopressin and oxytocin. Immunohistochemistry revealed patchy staining of medullary cells. The presence of vasopressin in the adrenal gland was also observed in Brattleboro rats.

In the syncytiotrophoblast of the *placenta,* Makino et al. (1983) found immunoreactive oxytocin. Fields et al. (1983) extracted immunoreactive and bioactive oxytocin from human placental tissue.

7 Functional Aspects of Extrahypothalamic Fiber Projections to Neural Target Areas

Vasopressin and oxytocin, when transported in the blood to distant targets, function as true hormones: they are bound to specific receptors in several target organs and tissues. At the level of the target cells they mediate very pronounced effects. Vasopressin causes antidiuresis by increasing the water permeability of the epithelial cells in the distal tubules of the kidney, or increase of blood pressure by contraction of vascular smooth muscle cells. Oxytocin causes uterine contraction and milk ejection by contraction of smooth muscle cells in the uterus and mammary gland.

The role of both peptides in various peripheral endocrine organs and other tissues where their occurrence has been described recently is yet unclear. Oxytocin, which is autochthonously produced in the ovary, appears to be involved in corpus luteum function and in steroidogenesis (Wathes et al. 1982).

What is the role of vasopressin and oxytocin transported in axons to neural target areas at the level of the effector neuron?

Cushing (1932) observed central effects of neurohypophyseal peptides after intraventricular injection of posterior pituitary exctract, which caused decrease of body temperature and vasodilatation. Vasopressin and oxytocin cause changes in the electrical activity of neurons in various parts of the nervous system (Gilberg et al. 1982; Morris et al. 1980; Mühlethaler et al. 1982). Electrophysiological studies in vivo in neural target areas of vasopressin and oxytocin fibers such as the locus coeruleus (Olpe and Baltzer 1981) and nucleus of the solitary tract (Morris et al. 1980) showed inducement of activation or inhibition of neurons. However, as was discussed by Buijs (1983), it is not certain that the neurons recorded are indeed innervated by vasopressin or oxytocin fibers. In vitro recording studies by

Table 4. Behavioral effects of arginine-vasopressin (AVP) and oxytocin (OXT) (Van Wimersma Greidanus 1984)

		AVP	OXT
Avoidance behavior			
Active avoidance	Acquisition	+	?
	Resistance to extinction	+	−
Passive avoidance	Consolidation	+	−
	Retrieval	+	−
Experimental amnesia	Prevention	+	0
	Reversal	+	0
Avoidance of attack		+	
Rewarded behavior			
Heroin self-administration	(Acquisition)	−	+
Electrical self-stimulation	(Threshold currents)	−	+
Sexual reward	(T-maze)	+	?
Sweet preference	(Maze performance)	+	?
Food reward	(Problem solving)	+	0/+
Thermoregulation			
Antipyretic		+	0
Cardiovascular regulation			
Centrally evoked pressor response		0/−	−
Maternal behavior	Facilitation	+/−	+
Other CNS effects			
Barrel rotation		+	0
Sexual behavior	(Lordosis)	−	?
Antinociceptive effects		+	?

Mühlethaler et al. (1982) showed excitation of hippocampal neurons by vasopressin which could be blocked by application of antagonists. Although these findings suggest the presence of a specific vasopressin antagonist, the study does not prove that these receptors are located in a neuron innervated by vasopressin terminals.

How are vasopressin and oxytocin inactivated at the site of the action? Burbach et al. (1980) demonstrated aminopeptidase activity capable of inactivating oxytocin in synaptic membrane fractions. Burbach et al. (1983) demonstrated that the degradation product of vasopressin, the hexapeptide ring and its desglycinamide, isolated from synaptic membranes is more effective than vasopressin in memory consolidation in a passive avoidance situation. Both peptides influence cyclic AMP levels (Schneider et al. 1982; Church 1983) and catecholamine turnover in certain specific brain areas (Versteeg et al. 1979). Van Wimersma Greidanus (1984) has recently compiled a survey of the various central functions of vasopressin and oxytocin (see Table 4). These include thermoregulation, cardiovascular regulation, antinociception, avoidance behavior, rewarded behavior, and other types of behavior.

Much interest has focussed on a wide variety of behavioral effects of exogenously applied vasopressin and oxytocin (De Wied 1977; De Wied and Versteeg

1979). These effects include extinction of avoidance response, retention of a passive avoidance response, prevention and reversal of experimentally induced amnesia (Flexner et al. 1977; Pfeiffer and Bookin 1978), learned helplessness (Leshner et al. 1978), maze learning, development of tolerance to and physical dependence on opiates (van Ree and DeWied 1977a, b; Moore 1983) and alcohol (Finkelberg et al. 1978; Hoffman et al. 1980; Myers et al. 1983; Rigter et al. 1980; Crabbe and Rigter 1980), drug-seeking behavior, sexual behavior (Södersten et al. 1983), maternal behavior (Rubin et al. 1983) and other types of behavior (see Table 4). From these observations it has been postulated that vasopressin and oxytocin affect memory processes and influence cognitive functions and that they oppositely induce storage and retrieval processes in the brain.

From the results of lesion studies and microinjection studies (Kovács et al. 1982) it was deduced that neurohypophyseal peptides affect memory processes by acting on mesolimbic structures (Van Wimersma Greidanus et al. 1983), especially in terminal regions of the coeruleotelencephalic noradrenaline system (Kovács et al. 1979a, b; Tanaka et al. 1977; Versteeg et al. 1978). Influence of vasopressin on dopamine in the amygdala or serotonin in the hippocampus has been described (Ramaekers, et al. 1977; Van Wimersma Greidanus and De Wied 1976; Van Wimersma Greidanus et al. 1975; Kovács et al. 1982). Van Wimersma Greidanus (1984) suggested that neurohypophyseal peptides have a physiological role on memory processes and cognitive functions based on (1) the observation of extrahypothalamic projections to limbic regions, (2) the pronounced effects of small centrally administered doses (De Wied 1976), (3) changes in cerebrospinal fluid levels in relation to behavioral performance, and (4) disturbed behavior of diabetes insipidus rats (De Wied et al. 1975) or of animals which had received vasopressin antiserum intraventricularly (De Wied 1976).

The effects of vasopressin on memory are dissociated from its peripheral effects on blood pressure and water metabolism (De Wied et al. 1984). Structure-activity relationship studies (Burbach et al. 1983) indicate that highly potent and selective memory-modulating peptide fragments can be formed of vasopressin and oxytocin.

Several studies have been carried out on the influence of neurohypophyseal peptides in normal human volunteers (Beckwith et al. 1982, 1983) and in patients with impaired memory due to various causes (Legros et al. 1978; Le Boeuf et al. 1978; Fewtrell et al. 1982; Laczi et al. 1983; Weingartner et al. 1981). In some studies improvements were reported. Unambiguous results may be obtained when methodology is improved (Jolles 1983).

Some controversy exists because the behavioral effects observed in the experimental model of the rat could not always be reproduced by other groups; the mode of action of vasopressin was explained differently (Le Moal et al. 1981; Gash and Thomas 1983; Saghal 1984). Danguir (1983) has postulated that the impaired learning performances of diabetes insipidus rats have to be attributed to disturbed paradoxical sleep (short sleep periods caused by frequent water intake).

Other central effects of vasopressin are increase of blood pressure (Pittman et al. 1982; Zerbe et al. 1983; Blessing et al. 1984; McNeill 1983; Montastruc et al. 1983; Matsuguchi et al. 1982), antipyresis (Kasting et al. 1980; Crine et al. 1981), and analgesia (Berkowitz and Sherman 1982; Berntson and Berson 1980; Kordower et al. 1982; Bodnar et al. 1980). Vasopressin and oxytocin are located in areas which participate in processes of autonomic regulation, nociception, and behavior.

In relation to autonomic regulation, the oxytocin and vasopressin fiber projections from the paraventricular nucleus to the neural lobe, portal vessels of the external zone of the median eminence, and autonomic centers in the brain stem (locus coeruleus, parabrachial nuclei, nucleus solitarius vagus complex) and spinal cord are of interest. The paraventricular nucleus, which consists of various subgroups of neurons receiving a great variety of different inputs, is involved in integration of endocrine and autonomic responses and in the regulation of the cardiovascular system (Sawchenko and Swanson 1982a, b), feeding behavior (Sawchenko 1982a, b), and thirst (Swanson and Mogenson 1981).

8 Summary

The neurophypophyseal hormones vasopressin and oxytocin are peptides with established peripheral endocrine effects. An increasing amount of data indicate that both peptides have distinct central effects on behavior, control of cardiovascular reflexes and temperature, pain modulation, etc. With immunohistochemical methods it has been demonstrated that vasopressin, oxytocin, and their accompanying neurophysins are transported from hypothalamic perikarya to fenestrated capillaries of the neural lobe as well as to a variety of nonhypophyseal vascular and neural target sites. Vasopressin and oxytocin are produced in magnocellular perikarya of the supraoptic nucleus and several accessory neurosecretory nuclei as well as in magnocellular and parvocellular perikarya of the paraventricular nucleus. In addition, vasopressin/neurphysin is produced in parvocellular perikarya of the suprachiasmatic nucleus, nucleus interstitialis striae terminalis, amygdala, dorsomedial hypothalamic nucleus, and locus coeruleus. In the paraventricular nucleus and to a lesser degree in the supraoptic nucleus, subsets of perikarya containing other neuropeptides (VIP, angiotensin II, cholecystokinin, opioid peptides, CRF, neurotensin) are found.

Coexistence of these peptides with vasopressin or oxytocin has been implicated, but at present, only coexistence of vasopressin and dynorphin and of oxytocin and cholecystokinin in some perikarya is likely. Vasopressin/neurophysin fibers terminating at portal vessels of the median eminence originate in the parvocellular part of the paraventricular nucleus and together with CRF fibers regulate ACTH secretion. Other vascular target areas are the organum vasculosum of the lamina terminalis, subfornical organ, and pineal. The functional role of vasopressin and oxytocin in these neurohemal circumventricular organs lacking a blood-brain barrier is yet unclear. Major neural target areas of vasopressin and/or oxytocin fibers and terminals are: olfactory bulb, frontal cortex, diagonal tract of Broca, nucleus accumbens, lateral septum, amygdala, nucleus interstitialis striae terminalis, dorsal and ventral hippocampus, lateral habenula, mediodorsal thalamus, periventricular nucleus, dorsomedial hypothalamus, posterior hypothalamus, nucleus supramamillaris, mesencephalic central gray, nucleus raphes dorsalis, substantia nigra, ventral tegmental area, nucleus interpeduncularis, nucleus parabrachialis dorsalis and ventralis, locus coeruleus, nucleus reticularis parvocellularis, nucleus raphes pontis, nucleus solitarius vagus complex, nucleus reticularis lateralis, nucleus raphes magnus and obscurus, substantia gelatinosa nervi trigemini, dorsal horn, central gray, and nucleus intermediolateralis of the spinal cord. Based on the combination of immunohistochemical neuropeptide staining with retro- and anterograde tract tracing methods, as well as on lesion studies, the following major fiber projections appear to be established. From the posterior parvocellular part of the paraventricular nucleus, vasopressin and oxytocin fibers project caudally to the mesencephalon, rhombencephalon and spinal cord. Projections from the suprachiasmatic nucleus terminate in the mediodorsal thalamus and dorsomedial and posterior hypothalamus. From parvocellular vasopressin/neurophysin perikarya of the nucleus interstitialis striae terminalis, fibers project to the lateral septum. Noteworthy are reports on the occurrence of vasopressin and oxytocin in several peripheral organs such as ovary, testis, and adrenals. The projections of vasopressin/neurophysin-containing fibers to several central sites, mainly in the mesolimbic system, brain stem, and spinal cord may be the structural correlate of central actions of these peptides and their fragments such as behavioral, cardiovascular, and antinociceptive effects.

References

Acher R (1980) Molecular evolution of biologically active polypeptides. Proc R Soc Lond [Biol] 210:21–43

Acher R, Manoussos G, Olivry G (1955) Sur les relations entre l'ocytocine et la vasopressine d'une part et la proteine de van Dyke d'autre part. Biochem Biophys Acta 16:155–156

Alonso G, Szafarczyk A, Assenmacher J (1984) Radioautographic evidence for extrahypothalamic efferences from the supraoptic nuclei in the rat. Proceed VIIth Int Congr Endocrinol Quebec July 1–7, Elsevier, Amsterdam, p 331

Alvarez-Buylla R, Livett BG, Uttenthal LO, Milton SH, Hope DB (1970) Immunohistochemical evidence for the transport of neurophysin in neurosecretory neurones of the dog. Acta Physiol Scand [Suppl] 357:5

Ananthanrazanan V (1955) Nature and distribution of neurosecretory cells of the reptilian brain. Z Zellforsch Mikroskop Anat 43:8–16

Antunes JL, Carmel PW, Zimmerman EA (1977) Projections from the paraventricular nucleus to the zona externa of the median eminence in the rhesus monkey: An immunohistochemical study. Brain Res 137:1–10

Armstrong DM, Pickel VM, Joh TH, Reis DJ, Miller RJ (1981) Immunocytochemical localization of catecholamine synthesizing enzymes and neuropeptides in area postrema and medial nucleus tractus solitarius of rat brain. J Comp Neurol 196:505–517

Armstrong DM, Pickel VM, Reis DJ (1982) Electron microscopic immunocytochemical localization of substance P in the area postrema of rat. Brain Res 243:141–146

Armstrong WE, Warach S, Hatton GI, McNeill TH (1980) Subnuclei in the rat hypothalamic paraventricular nucleus: A cytoarchitectural, horseradish peroxidase and immunocytochemical analysis. Neuroscience 5:1931–1958

Armstrong WE, Schöler J, McNeill TH (1982) Immunocytochemical, Golgi and electron microscopic characterization of putative dendrites in the ventral glial lamina of the rat supraoptic nucleus. Neuroscience 7:679–694

Bargmann W (1949) Über die neurosekretorische Verknüpfung von Hypothalamus and Neurohypophyse. Z Zellforsch Mikroskop Anat 34:610–634

Barnard RR jr, Morris M (1982) Cerebrospinal fluid vasopressin and oxytocin: evidence for an osmotic response. Neurosci Lett 29:275–279

Barnes KL, Ferrario CM, Conomy JP (1979) Comparison of the hemodynamic changes produced by electrical stimulation of the area postrema and nucleus tractus solitarii in the dog. Circ Res 45:136–143

Barry J (1956) Les voies extra-hypophysaires de la neurosécrétion diencephalique. Bull Assoc Anat (Nancy) 89:264–276

Barry J (1961) Recherches morphologiques et expérimentales sur la glande diencéphalique et l'appareil hypothalamo-hypophysaire. Ann Scientif Univers Besancon, Zool Physiol 2:3–133

Beckwith BE, Petros T, Kanaan-Beckwith S, Couk DI, Haug RJ (1982) Vasopressin analog (DDAVP) facilitates concept learning in human males. Peptides 3:627–630

Beckwith BE, Couk DI, Till TS (1983) Vasopressin analog influences the performance of males on a reaction time task. Peptides 4:707–709

Beinfeld MC, Meyer DK, Brownstein MJ (1980) Cholecystokinin octapeptide in rat hypothalamo-neurohypophyseal system. Nature 288:376–378

Berk ML, Reaves TA jr, Hayward JN, Finkelstein JA (1982) The localization of vasotocin and neurophysin neurones in the diencephalon of the pigeon *Columba livia*. J Comp Neurol 204:392–406

Berkowitz BA, Sherman S (1982) Characterization of vasopressin analgesia. J Pharmacol Exp Ther 220:329–334

Berntson GG, Berson BS (1980) Antinociceptive effects of intraventricular or systemic administration of vasopressin in the rat. Life Sci 26:455–459

Blessing WW, Chalmers JP (1979) Direct projection of catecholamine (presumably dopamine) containing neurons from hypothalamus to spinal cord. Neurosci Lett 11:35–40

Blessing WW, Sved AF, Reis DJ (1984) Arterial pressure and plasma vasopressin: Regulation by neurons in the caudal ventrolateral medulla of the rabbit. Clin Exp Hypertens [A]6:149–156

Blume HW, Pittmann GJ, Renaud LP (1978) Electrophysiological indications of a "vasopressinergic" innervation of the median eminence. Brain Res 155:153–158

Bodnar RJ, Zimmerman EA, Nilaver G, Mansour A, Thomas LW, Kelly DD, Glusman M (1980) Dissociation of cold-water swim and morphine analgesia in Brattleboro rats with diabetes insipidus. Life Sci 26:1581–1590

Boer GJ, van Rheenen-Verberg CMH, Uylings HBM (1982) Impaired brain development of the diabetes insipidus Brattleboro rat. Dev Brain Res 3:557–575

Bonner TI, Brownstein MJ (1984) Vasopressin, tissue-specific defects and the Brattleboro rat. Nature 310:17

Bons N (1980a) Charactérisation immunocytochimique des systémes neurosécréteurs à mesotocine et à vasotocine dans l'encéphale du Canard. C R Seances Acad Sc III 290:113–116

Bons N (1980b) The topography of mesotocin and vasotocin systems in the brain of the domestic mallard and Japanese quail: Immunocytochemical identification. Cell Tissue Res 213:37–51

Bons N (1983) Immunocytochemical identification of the mesotocin- and vasotocin-producing systems in the brain of temperate and desert lizard species and their modifications by cold exposure. Gen Comp Endocrinol 52:56–66

Bouchaud C (1979) Evidence for a multiple innervation of subcommissural ependymocytes in the rat. Neurosci Lett 12:253–258

Bronzino JD, Morgane PJ, Stern WC (1972) EEG synchronization following application of serotonin to area postrema. Am J Physiol 223:376–383

Brownfield MS, Kozlowski GP (1977) The hypothalamo-choroidal tract I. Immunohistochemical demonstration of neurophysin pathways to telencephalic choroid plexuses and cerebrospinal fluid. Cell Tissue Res 178:111–127

Bugnon C, Fellmann D, Gouget A (1983) Changes in corticoliberin and vasopressin-like immunoreactivities in the zona externa of the median eminence in adrenalectomized rats. Immunocytochemical study. Neurosci Lett 37:43–49
Buijs RM (1978) Intra- and extrahypothalamic vasopressin and oxytocin pathways in the rat. Pathways to the limbic system, medulla oblongata and spinal cord. Cell Tissue Res 192:423–435
Buijs RM (1983) Vasopressin and oxytocin – their role in neurotransmission. Pharmacol Ther 22:127–141
Buijs RM, Pévet P (1980) Vasopressin and oxytocin containing fibres in the pineal gland and subcommissural organ of the rat. Cell Tissue Res 205:11–17
Buijs RM, Swaab DF (1979) Immunoelectron microscopical demonstration of vasopressin and oxytocin synapses in the rat limbic system. Cell Tissue Res 204:355–365
Buijs RM, Swaab DF, Dogterom J, Van Leeuwen FW (1978) Intra- and extrahypothalamic vasopressin and oxytocin pathways in the rat. Cell Tissue Res 186:423–433
Buijs RM, van Heerikhuize JJ (1982) Vasopressin and oxytocin release in the brain: A synaptic event. Brain Res 252:71–76
Buijs RM, Velis DN, Swaab DF (1980a) Extrahypothalamic vasopressin and oxytocin innervation of fetal and adult rat brain. In: Adaptive capabilities of the nervous system. Prog Brain Res 53:159–167
Buijs RM, Velis DN, Swaab DF (1980b) Ontogeny of vasopressin and oxytocin in the fetal rat: Early vasopressinergic innervation of the fetal brain. Peptides 1:315–324
Burbach JPH, De Kloet ER, De Wied D (1980) Oxytocin biotransformation in the rat limbic brain: characterization of peptidase activities and significance in the formation of oxytocin fragments. Brain Res 202:401–414
Burbach JPH, Kovács GL, De Wied D, van Nispen JW, Greven HM (1983) A major metabolite of arginine vasopressin in the brain is a highly potent neuropeptide. Science 221:1310–1312
Burlet A, Chateau M, Czernichow P (1979) Immunocytochemical study of neurohypophysial peptides during corticotropic maturation of infant rats. Cell Tissue Res 201:315–325
Burlet A, Tonon MC, Tankosic P, Coy D, Vaudry H (1983) Comparative immunocytochemical localization of corticotropin releasing factor (CRF-41) and neurohypophyseal peptides in the brain of Brattleboro and Long-evans rats. Neuroendocrinology 37:64–72
Caffé AR, Van Leeuwen FW (1983) Vasopressin-immunoreactive cells in the dorsomedial hypothalamic region, medial amygdaloid nucleus and locus coeruleus of the rat. Cell Tissue Res 233:23–33
Camacho A, Philipps IM (1981) Horseradish peroxidase study in rat of the neural connections of the organum vasculosum of the lamina terminalis. Neurosci Lett 25:201–204
Card JP, Moore RY (1982) Ventral lateral geniculate nucleus efferents to the rat suprachiasmatic nucleus exhibit avian pancreatic polypeptide-like immunoreactivity. J Comp Neurol 206:390–396
Card JP, Brecha N, Karten HJ, Moore RY (1981) Immunocytochemical localization of vasoactive instestinal polypeptide containing cells and processes in the suprachiasmatic nucleus of the rat. Light and electron microscopic analysis. J Neurosci 1:1289–1303
Choy VS, Watkins WB (1979) Maturation of the hypothalamo-neurohypophysial system. I. Localization of neurophysin, oxytocin and vasopressin in the hypothalamus and neural lobe of the developing rat brain. Cell Tissue Res 197:325–336
Church AC (1983) Vasopressin potentiates the stimulation of cyclic AMP accumulation by norepinephrine. Peptides 4:261–263
Coghlan JP, Penschow JD, Hudson PJ, Niall HD (1984) Hybridization histochemistry: use of recombinant DNA for tissue localizations of specific mRNA populations. Clin Exp Hypertens [A] 6:63–78
Conrad LCA, Pfaff DW (1976) Efferents from medial basal forebrain and hypothalamus of the rat. II. An autoradiographic study of the anterior hypothalamus. J Comp Neurol 169:221–262
Coons AH (1956) Histochemistry with labeled antibody. Int Rev Cytol 5:1–23

Coons AH (1958) Fluorescent methods. In: Danielli JF (ed) General cytochemical methods. Academic Press, New York, pp 399–422
Coulter HD, Elde RP, Unverzagt SL (1981) Co-localization of neurophysin-like and enkephalin-like immunoreactivity in cat pituitary. Peptides 2 [Suppl] 1:51–55
Crabbe JC, Rigter H (1980) Learning and the development of alcoholtolerance and dependence. The role of vasopressin-like peptides. Trends Neurosci:20–23
Crabbe JC, Rigter H (1980) Learning and the development of alcohol tolerance and depenways in the North American Opossum (*Didelphys virginiana*). Studies using the horseradish peroxidase method. J Comp Neurol 179:169–194
Crine AF, Bredart S, Legros JJ (1981) Effects of exogenous arginine vasopressin on rectal temperature in the albino rat. Hormones Behav 15:226–231
Cushing H (1932) Paper relating to the pituitary body. In: Hypothalamus and parasympathetic nervous system. Thomas, Springfield
Danguir J (1983) Sleep deficits in rats with hereditary diabetes insipidus. Nature 304:163–164
De Vries GJ, Buijs RM (1983) The origin of the vasopressinergic and oxytocinergic innervation of the rat brain, with special reference to the lateral septum. Brain Res 273:307–317
De Vries GJ, Buijs RM, Swaab DF (1981) Ontogeny of the vasopressinergic neurons of the nucleus suprachiasmaticus and their extrahypothalamic projections in the rat brain. Presence of a sex difference in the lateral septum. Brain Res 218:67–78
De Wied D (1976) Behavioral effects of intraventricularly administered vasopressin and vasopressin fragments. Life Sci 19:685–690
De Wied D (1977) Peptides and behavior. Life Sci 20:195–204
De Wied D, Versteeg DHG (1979) Neurohypophyseal principles and memory. Fed Proc 38:2348–2354
De Wied D, Bohus B, van Wimersma Greidanus TJB (1975) Memory deficit in rats with hereditary diabetes insipidus. Brain Res 85:152–156
De Wied D, Gaffori O, van Ree JM, de Jong W (1984) Central target for the behavioural effects of vasopressin neuropeptides. Nature 308:276–278
Dierickx K (1980) Immunocytochemical localization of the vertebrate cyclic nonapetide neurohypophyseal hormones and neurophysins. Int Rev Cytol 62:119–185
Dierickx K, Vandesande F (1977) Immunocytochemical demonstration in the external region of the amphibian median eminence of seperate vasotocinergic and mesotocinergic nerve fibers. Cell Tissue Res 177:47–56
Dierickx K, Vandesande F (1979a) Immunocytochemical demonstration of separate vasopressin-neurophysin and oxytocin-neurophysin neurons in the human hypothalamus. Cell Tissue Res 196:203–212
Dierickx K, Vandesande F (1979b) Immunocytochemical localization of somatostatin neurons in the rat hypothalamus. Cell Tissue Res 201:349–359
Dierickx K, Vandesande F, De Mey J (1976) Identification in the external region of the rat median eminence of separate neurophysin-vasopressin and neurophysin-oxytocin containing nerve fibres. Cell Tissue Res 168:141–151
Dogterom J, van Wimersma Greidanus TB, Swaab DF (1977) Evidence for the release of vasopressin and oxytocin into cerebrospinal fluid: measurements in plasma and CSF of intact and hypophysectomized rats. Neuroendocrinology 24:108–118
Dogterom J, Snijdewint FGM, Buijs RM (1978a) The distribution of vasopressin and oxytocin in the rat brain. Neurosci Lett 9:341–346
Dogterom J, van Wimersma Greidanus TB, De Wied D (1978b) Vasopressin in cerebrospinal fluid and plasma of man, dog and rat. Am J Physiol 234:E463–E467
Dogterom J, Snijdewint FGM, Pévet P, Swaab DF (1980) Studies on the presence of vasopressin, oxytocin and vasotocin in the pineal gland, subcommissural organ and fetal pituitary gland: failure to demonstrate vasotocin in mammals. J Endocrinol 84:115–123
Dorsa DM, Bottemiller L (1982) Age related changes of vasopressin content of microdissected areas of the rat brain. Brain Res 242:151–156

Dorsa DM, Majumdar LA, Petracca FM, Baskin DG, Cornett LE (1983) Characterization and localization of ^{3}H-arginine^8-vasopressin binding to rat kidney and brain tissue. Peptides 4:699–706
Du Vigneaud V (1954) Hormones of the posterior pituitary gland: oxytocin and vasopressin. Harvey Lect 50:1–26
Dyball REJ, Kemplay SK (1982) Dendritic trees of neurones in the rat supraoptic nucleus. Neuroscience 7:223–230
Edwards GL, Ritter RC (1981) Ablation of the area postrema causes exaggerated consumption of preferred foods in the rat. Brain Res 216:265–276
Ellis HK, Watkins WB (1975) Ontogeny of the pig hypothalamic neurosecretory system with particular reference to the distribution of neurophysin. Cell Tissue Res 164:543–557
Felix D, Akert K (1974) The effect of angiotensin II on neurones of the cat subfornical organ. Brain Res 169:204–208
Ferrario CM, Barnes KL, Szilagyi JE, Brosnihan KB (1979) Physiological and pharmacological characterization of the area postrema pressor pathways in the normal dog. Hypertension 1:235–245
Fewtrell WD, House AO, Jamie PF, Oates MR, Cooper JE (1982) Effects of vasopressin on memory and new learning in a brain-injured population. Psychosom Med 12:423–425
Fields PhA, Eldridge RK, Fuchs A-R, Roberts RF, Fields MJ (1983) Human placental and bovine corpora luteal oxytocin. Endocrinology 112:1544–1546
Finkelberg F, Kalant H, Le Blanc AE (1978) Effect of vasopressin-like peptides on consumption of ethanol by the rat. Pharmacol Biochem Behav 9:453–458
Fisher AWF, Price PG, Burford GD, Lederis K (1979) A 3-dimensional reconstruction of the hypothalamo-neurohypophysial system of the rat. The neurons projecting to the neuro/intermediate lobe and those containing vasopressin and somatostatin. Cell Tissue Res 204:343–354
Flexner JB, Flexner LB, Hoffman PL, Walter R (1977) Dose-response relationships in attenuation of puromycin-induced amnesia by neurohypophyseal peptides. Brain Res 134:139–144
Flint APF, Sheldrick EL (1983) Evidence for a systemic role for ovarian oxytocin in luteal regressive sheep. J Reprod Fertil 67:215–225
Fuxe K (1965) Evidence for the existence of monoamine neurons in the central nervous system. IV. Distribution of monoamine nerve terminals in the central nervous system. Acta Physiol Scand 64 [Suppl] 247:37–85
Fuxe K, Agnati LF, Ganten D, Lang RE, Calza L, Poulsen K, Infantinellina F (1982) Morphometric evaluation of the coexistence of renin-like and oxytocin-like immunoreactivity in nerve cells of the paraventricular hypothalamic nucleus of the rat. Neurosci Lett 33:19–24
Gainer H, Sarne Y, Brownstein MJ (1977) Biosynthesis and axonal transport of rat neurohypophysial proteins and peptides. J Cell Biol 73:366–381
Ganten D, Fuxe K, Phillips IM, Mann JFE, Ganten U (1978) The brain isorenin-angiotensin system: biochemistry, localization, and possible role in drinking and blood pressure regulation. In: Ganong WF, Martini L (eds) Frontiers in Neuroendocrinology 5 Raven Press, New York, pp 61–99
Gash DM, Thomas GJ (1983) What is the importance of vasopressin in memory processes? Trends Neurosci 6:197–198
Gash D, Sladek C, Scott D (1980) Cytodifferentiation of the supraoptic nucleus correlated with vasopressin synthesis in the rat. Brain Res 181:345–355
Gaupp R (1944) Ein weiterer Beitrag zur pathologischen Anatomie des Diabetes insipidus. Z Ges Neurol Psychiatr 177:50–73
Gauquelin G, Geelen G, Louis F, Allevard AM, Meunier C, Cuisinaud G, Benjanet S, Seidah NG, Chrétien M, Legros JJ, Gharib C (1983) Presence of vasopressin, oxytocin and neurophysin in the retina of mammals, effect of light and darkness, comparison with the neuropeptide content of the neurohypophysis and the pineal gland. Peptides 4:509–515

George IM, Jacobowitz DM (1975) Localization of vasopressin in discrete areas of rat hypothalamus. Brain Res 93:363–366

Gilberg MP, Cooke JH, Fleetwood-Walker S, Peterson DF (1982) The influence of the paraventriculo-spinal pathway and oxytocin and vasopressin on sympathetic preganglionic neurons. Brain Res 251:283–290

Girardie J, Remy C (1980) Particularités histo-cytologiques des prolongements distaux de 2 cellules à „vasopressine-neurophysinelike" du Criquet migrateur. J Physiol (Paris) 76:265–271

Glick SM, Brownstein M (1980) Vasopressin content of rat brain. Life Sci 27:1103–1110

Gold PW, Weingartner H, Ballenger JC, Goodwin FK, Post RM (1979) Effects of 1-desamino-8-D-arginine vasopressin on behavior and cognition in primary affective disorder. Lancet 2:992–994

Gomori G (1941) Observations with differential stains on human islets of Langerhans. Am J Pathol 17:395–406

Goossens N, Dierickx K, Vandesande F (1977 a) Immunocytochemical demonstration of the hypothalamo-hypophysial vasotocinergic system of *Lampetra fluviatilis*. Cell Tissue Res 177:317–323

Goossens N, Dierickx K, Vandesande F (1977 b) Immunocytochemical localization of vasotocin and isotocin in the preoptico-neurohypophysial system of teleost. Gen Comp Endocrinol 32:371–375

Goossens N, Dierickx K, Vandesande F (1978) Immunocytochemical study of the neurohypophysial hormone producing system of the lungfish, *Protopterus aethiopicus*. Cell Tissue Res 190:69–77

Goosens N, Dierickx K, Vandesande F (1979) Immunocytochemical localization of vasotocin and mesotocin in the hypothalamus of lacertilian reptiles. Cell Tissue Res 200:223–227

Grimmelikhuijzen CJP, Dierickx K, Boer GJ (1982) Oxytocin/vasopressinlike immunoreactivity is present in the nervous system of hydra. Neurosci 7:3191–3199

Guldenaar SEF, Wathes DC, Pickering BT (1984) Immunocytochemical evidence for the presence of oxytocin and neurophysin in the large cells of the bovine corpus luteum. Cell Tissue Res 237:349–352

Hancock MR (1976) Cells of origin of hypothalamo-spinal projections in the rat. Neurosci Lett 3:179–184

Hanley MR, Benton HP, Lightman SL, Todd K, Bone EA, Fretten P, Palmer S, Kirk CJ, Michell RH (1984) A vasopressin-like peptide in the mammalian sympathetic nervous system. Nature 309:258–261

Hawthorn J, Ang VTY, Jenkins JS (1980) Localisation of vasopressin in the rat brain. Brain Res 197:75–81

Hawthorn J, Graham JM, Jenkins JS (1984) Localization of vasopressin in synaptic vesicles of extrahypothalamic rat brain. Life Sci 35:277–284

Haywood JR, Fink GD, Buggy J, Phillips MI, Brody MJ (1980) The area postrema plays no role in the pressor action of angiotensin in the rat. Am J Physiol 239 (Heart Circ Physiol 8): H108–H113

Heller H, Hasan SH, Saifi AQ (1968) Antidiurectic activity in the cerebrospinal fluid. J Endocrinology 41:273–280

Hoffman PL, Ritzmann RF, Tabakoff B (1980) Neurohypophyseal peptide influences on ethanol tolerance and acute effects of ethanol. Pharmacol Biochem Behav 13 [Suppl 1]: 279–284

Hoorneman EMD, Buijs RM (1982) Vasopressin fiber pathways in the rat brain following suprachiasmatic nucleus lesioning. Brain Res 243:235–241

Hosoya Y, Matsushita M (1979) Identification and distribution of the spinal and hypophyseal projection neurons in the paraventricular nucleus of the rat. A light and electron microscopic study with the horseradish peroxidase method. Exp Brain Res 35:315–331

Hosoya Y, Matsushita M (1981) A direct projection from the hypothalamus to the area postrema in the rat, as demonstrated by the HRP and autoradiographic methods. Brain Res 214:144–149

Hyde TM, Miselis RR (1983) Effects of area postrema/caudal medial nucleus of solitary tract lesions on food intake and body weight. Am J Physiol 244 (Reg Integr Comp Physiol 13):R577–R587

Jenkins JS, Mather HM, Ang VTY (1980) Vasopressin in human cerebrospinal fluid. J Clin Endocr Metab 50:364–367

Jenkins JS, Ang VTY, Hawthorn J, Rossor MN, Iversen LL (1984) Vasopressin, oxytocin and neurophysins in the human brain and spinal cord. Brain Res 291:111–117

Jolles J (1983) Vasopressin-like peptides and the treatment of memory disorders in man. In: The neurohypophysis: Structure, function and control. Prog Brain Res 60:169–182

Kahn P, Abrams GM, Zimmerman E, Carraway R, Leeman S (1980) Neurotensin neurons in the rat hypothalamus. An immunocytochemical study. Endocrinology 107:47–51

Kalia M, Fuxe K, Hökfelt T, Johansson T, Lang R, Ganten D, Cuello C, Terenius L (1984) Distribution of neuropeptide immunoreactive nerve terminals within the subnuclei of the tractus solitarius of the rat. J Comp Neurol 222:409–444

Kasson G, Meidan R, Hsueh AJW (1984) Identification of vasopressinlike substances in the rat testis. Proc VIIth Int Congr Endocrinol, Quebec Juli 1–7, Elsevier, Amsterdam. p 938

Kasting NW, Veale WL, Cooper KE (1980) Vasopressin: a homeostatic effector in the febrile process. Neurosci Biol Rev 6:215–222

Kawata M, Hashimoto K, Takahara J, Sano Y (1983a) Differences in the distributional pattern of CRF-, oxytocin-, and vasopressin-immunoreactive nerve fibers in the median eminence of the rat. Cell Tissue Res 230:247–258

Kawata M, Ueda S, Sano Y (1983b) Two types of oxytocin and vasopressin nerve fibers in the intra- and extrahypothalamic neuronal systems as revealed by immunohistochemistry. Acta Anat (Basel) 116:193–200

Kelly J, Swanson LW (1980) Additional forebrain regions projecting to the posterior pituitary: preoptic region, bed nucleus of the stria terminalis and zona incerta. Brain Res 197:1–9

Keppens S, De Wulf H (1979) The nature of the hepatic receptors involved in vasopressin-induced glycogenolysis. Biochem Biophys Acta 538:63–69

Khan-Dawood FS, Dawood MY (1984) Oxytocin content of human fetal pituitary glands. Am J Obstet Gynecol 4:420–423

Khan-Dawood FS, Marut EL, Dawood MY (1984) Oxytocin in the corpus luteum of the cynomolgus monkey (*Macaca fascicularis*). Endocrinology 115:570–574

Kilcoyne M, Hoffman DL, Zimmerman EA (1980) Immunocytochemical localization of angiotensin II and vasopressin in rat hypothalamus: evidence for production in the same neuron. Clin Sci 59:57s–60s

Kiss JZ, Williams TH, Palkovits M (1984) Distributions and projections of cholecystokinin immunoreactive neurons in the hypothalamic paraventricular nucleus of the rat. J Comp Neurol 227:173–181

Kneisley LW, Biber MP, La Vail JH (1978) A study of the origin of brainstem projections to monkey spinal cord using the retrograde transport method. Exp Neurol 60:116–139

Kok TP, van der Sluis PJ, Boer GJ (1983) Chemical identification of the vasopressin immunoreactive material present in the rat suprachiasmatic nucleus. Neuropeptides 3:255–262

Kordower JH, Sikorszky V, Bodnar RJ (1982) Central antinociceptive effects of lysine-vasopressin and an analogue. Peptides 3:613–617

Korf HW, Møller M (1984) The innervation of the mammalian pineal gland with special reference to central pinealopetal projections. Pineal Res Rev 2:41–86

Kovács GL, Bohus B, Versteeg DHG (1979a) Facilitation of memory consolidation by vasopressin: mediation by terminals of the dorsal noradrenergic bundle? Brain Res 172:73–85

Kovács GL, Bohus B, Versteeg DHG (1979b) The effects of vasopressin on memory processes: the role of noradrenergic neurotransmission. Neuroscience 4:1529–1537

Kovács GL, Buijs RM, Bohus B, van Wimersma Greidanus TB (1982) Microinjection of arginine^{8}-vasopressin antiserum into the dorsal hippocampus attenuates passive avoidance behavior in rats. Physiol Behav 28:45–48

Kozlowski GP, Nilaver G, Zimmerman EA (1983) Distribution of neurohypophysial hormones in the brain. Pharmacol Ther 21:325–349
Krisch B (1980) Electron microscopic immunocytochemical investigation on the postnatal development of the vasopressin system. Cell Tissue Res 205:453–471
Kuypers HGJM, Maisky VA (1975) Retrograde axonal transport of horseradish peroxidase from spinal cord to brainstem cell groups in the cat. Neurosci Lett 1:9–14
Laczi F, Van Ree JM, Balogh L, Szász A, Járdánházy T, Wágner A, Gáspár L, Valkusz Z, Dobranovics I, Szilárd J, Lázló FA, De Wied D (1983) Lack of effect of desglycinamide-arginine-vasopressin (DGAVP) on memory in patients with Korsakoff's syndrome. Acta Endocrinol (Copenh) 104:177–182
Land H, Schütz G, Schmale H, Richter D (1982) Nucleotide sequence of cloned cDNA encording bovine arginine vasopressin-neurophysin II precursor. Nature 295:299–303
Land H, Grez M, Ruppert S, Schmale H, Rehbein M, Richter D, Schütz G (1983) Deduced amino acid sequence from the bovine oxytocin-neurophysin I cDNA. Nature 302:342–344
Le Boeuf A, Lodge J, Eames PG (1978) Vasopressin and memory in Korsakoff syndrome. Lancet:1370
Lechan RM, Nestler JL, Jacobson S, Reichlin S (1980) The hypothalamo-tuberoinfundibular system of the rat as demonstrated by horseradish peroxidase (HRP) microiontophoresis. Brain Res 195:13–27
Legait H (1957) Anatomie microscopique des noyaux hypothalamiques neurosécrétoires et de leurs voies efferentes chez la Poule Rhode Island. Acta Neuroveg 15:252–262
Legait H, Legait E (1957) Les voies extrahypophysaires des noyaux neurosécrétoires hypothalamiques chez les batraciens et les reptiles. Acta Anat (Basel) 30:429–443
Legros JJ, Gilot P, Seron X, Claessens J, Adam A, Moeglen JM, Audibert A, Berchier P (1978) Influence of vasopressin on learning and memory. Lancet:41–42
Le Moal M, Koob GF, Koda LJ, Bloom FE, Manning M, Sawyer WH, Rivier J (1981) Vasopressor receptor antagonist prevents behavioral effects of vasopressin. Nature 291:491–493
Leshner AI, Hofstein R, Samuel D, van Wimersma Greidanus TB (1978) Intraventricular injection of antivasopressin serum blocks learned helplessness in rats. Pharmacol Biochem Behav 9:889–892 (1978)
Lim ATW, Lolait SJ, Barlow JW, Autelitano DJ, Toh BH, Boublik J, Abraham J, Johnston CI, Funder JW (1984) Immunoreactive arginine-vasopressin in Brattleboro rat ovary. Nature 310:61–64
Lind RW, Ohman LE, Lansing MB, Johnson AK (1983) Transection of subfornical organ neural connections diminishes the pressor response to intravenously infused angiotensin II. Brain Res 275:361–364
Lindvall O, Björklund A (1974) The organization of the ascending catecholamine neurons system in the rat brain, as revealed by the glyoxylic acid fluorescence method. Acta Physiol Scand [Suppl] 412:1–48
Lorén I, Alumets J, Håkanson R, Sundler F (1979) Distribution of gastrin and CCK-like peptides in rat brain. Histochemistry 59:249–258
Lu CL, Cantin M, Seidah NG, Chrétien M (1982) Distribution pattern in the human pituitary and hypothalamus of a new neuropeptide: the C-terminal glycoprotein-fragment of human pro-pressophysin (CPP). Histochemistry 75:319–326
Luerssen TG, Robertson GL (1980) Cerebrospinal fluid vasopressin and vasotocin in health and disease. In: Wood JH (ed) Neurobiology of cerebrospinal fluid vol 1. Plenum, New York, pp 613–623
Lutz-Bucher B, Koch B (1983) Characterization of specific receptors for vasopressin in the pituitary gland. Biochem Biophys Res Commun 115:492–498
Makara GB, Stark E, Karteszi M, Palkovits M, Rappay GY (1981) Effects of paraventricular lesions on stimulated ACTH release and CRF in stalk-median eminence of the rat. J Physiol 240:E441–446
Makino T, Nakazawa K, Ishii K, Haginiwa I, Nakayama A, Iizuka R (1983) Detection of immunoreactive human placental oxytocin and its contractile effect on the uterine muscle. Endocrinol Jpn 30:389–395

Mannucci PM, Aberg DM, Nilsson BJM, Robertson B (1975) Mechanism of plasminogen activation and factor VIII increase after vasoactive drugs. Br J Haematol 30:81
Martin G, Baverel G (1981) Vasopressin promotes the metabolism of near-physiological concentration of glutamine in isolated rat liver cells. Biosci Rep 4:171
Martin R, Voigt KH (1981) Enkephalins co-exist with oxytocin and vasopressin in nerve terminals of rat neurohypophysis. Nature 289:502–504
Martin R, Geis R, Holl R, Schäfer M, Voigt KH (1983a) Co-existence of unrelated peptides in oxytocin and vasopressin terminals of rat neurohypophysis: immunoreactive methionine^5-enkephalin, leucine^5-enkephalin- and cholecystokinin-like substances. Neuroscience 8:213–227
Martin R, Moll U, Voigt KH (1983b) An attempt to characterize by immunocytochemical methods the enkephalin-like material in oxytocin endings of the rat neurohypophysis. Life Sci 33:69–72
Mason WT, Ho YW, Eckenstein F, Hatton GI (1983) Mapping of cholinergic neurones associated with rat supraoptic nucleus: combined immunocytochemical and histochemical identification. Brain Res Bull 11:617–626
Mason WT, Ho YW, Hatton GI (1984) Axon collaterals of supraoptic neurons: Anatomical and electrophysiological evidence for their existence in the lateral hypothalamus. Neuroscience 11:169–182
Matsuguchi H, Sharabi FM, Gordon FJ, Johnson AK, Schmid PG (1982) Blood pressure and heart rate responses to microinjection of vasopressin into the nucleus tractus solitarius region of the rat. Neuropharmacology 21:687–693
Matsuura T, Kawata M, Yamada H, Kojima M, Sano Y (1983) Immunohistochemical studies on the peptidergic nerve fibres in the pineal organ of the dog. Arch Histol Jp 46:373–379
McGlone S, Ritter JJ, Kelley KW (1980) The antiaggressive effect of lithium is abolished by area postrema lesions. Physiol Behav 24:1095–1100
McKellar S, Loewy AD (1981) Organization of some brain stem afferents to the paraventricular nucleus of the hypothalamus of the rat. Brain Res 217:351–357
McNeill JR (1983) Role of vasopressin in the control of arterial pressure. Can J Physiol Pharmacol 61:1226–1235
McNeill TH, Kozlowski GP, Abel JH jr, Zimmerman EA (1976) Neurosecretory pathways in the mallard duck (*Anas platyrhynchos*) brain: localization by aldehyde fuchsin and immunoperoxidase techniques for neurophysin (NP) and gonadotrophin releasing hormone. Endocrinology 99:1323–1332
Mens WBJ (1983) Neurohypophyseal hormones in blood, cerebrospinal fluid and brain of the rat. Pharm Weekbl [Sci] 5:79–80
Mens WBJ, Andringa-Bakker EAD, van Wimersma Greidanus TB (1982) Changes in cerebrospinal fluid levels of vasopressin and oxytocin of the rat during various light-dark regimes. Neurosci Lett 34:51–56
Meyer DK, Oertel WH, Brownstein MJ (1980) Deafferentation studies on the glutamic acid decarboxylase content of the supraoptic nucleus of the rat. Brain Res 200:165–168
Micevych P, Elde R (1980) Relationship between enkephalinergic neurons and the vasopressin-oxytocin neuroendocrine system of the cat: an immunohistochemical study. J Comp Neurol 190:135–146
Michelini LC, Barnes KL, Ferrario CM (1983) Arginine vasopressin modulates the central action of angiotensin II in the dog. Hypertension [Suppl 5] 5:V94–V100
Miselis R (1981) The efferent projections of the subfornical organ of the rat; a circumventricular organ within a neural network subserving water balance. Brain Res 230:1–23
Miselis R (1982) The subfornical organ's neural connections and their role in water balance. Peptides 3:501–502
Møllgard K, Wiklund L (1979) Serotonergic synapses on ependymal and hypendymal cells of the rat subcommissural organ. J Neurocytol 8:445–467
Montastruc JL, Dang Tran L, Montastruc P (1983) Peptides neuro-hypophysaires et contrôle central cardiovasculaire. Arch Mal Coeur 76:9–12 (1983)

Moore JE (1983) Arginine vasopressin enhances retention of morphine tolerance. Pharmacol Biochem Behav 19:561–565
Moore RY (1973) Retinohypothalamic projection in mammals. A comparative study. Brain Res 49:403–409
Moore RY, Halaris AE, Jones BE (1978) Serotonin neurons of the midbrain raphe: Ascending projections. J Comp Neurol 180:417–438
Moore RY, Gustafson EL, Card PJ (1984) Identical immunoreactivity of afferents to the rat suprachiasmatic nucleus with antisera against avian pancreatic polypeptide, molluscan cardioexcitatory peptide and neuropeptide Y. Cell Tissue Res 236:41–46
Morris R, Salt THE, Sofroniew MV, Hill RG (1980) Actions of microintophoretically applied oxytocin and immunohistochemical localization of oxytocin, vasopressin and neurophysin in the rat caudal medulla. Neurosci Lett 18:163–168
Mühlethaler M, Dreifuss JJ, Gähwiler BH (1982) Vasopressin excites hippocampal neurones. Nature 296:749–751
Mühlethaler M, Raggenbass M, Dreifuss J (1984) Peptides related to vasopressin in invertebrates. Experientia 40:777–782
Myers RD, Critcher EC, Cornwell NN (1983) Effect of chronic vasopressin treatment on alcohol drinking of Brattleboro HZ and DI rats. Peptides 4:359–366
Namboodiri MAA, Favilla JT, Klein DC (1981) Pineal N-acetyltransferase is inactivated by disulfide containing peptides: Insulin is the most potent. Science 213:571–573
Negro-Vilar A, Sanchez-Franco F, Kwiatkowski M, Samson WK (1979) Failure to detect radioimmunoassayable arginine vasotocin in mammalian pineals. Brain Res Bull 4:789–792
Nicholson HD, Swann RW, Burford GD, Wathes DC, Porter DG, Pickering BT (1984) Identification of oxytocin and vasopressin in the testis and in adrenal tissue. Regul Pept 8:1141–1146
Nilaver G, Zimmerman EA, Wilkins J, Michaels J, Hoffmann D, Silverman AJ (1980) Magnocellular hypothalamic projections to the lower brain stem and spinal cord of the rat. Neuroendocrinology 30:150–158
Nilaver G, Mulhern J, Zimmerman EA (1982) Extrahypothalamic neurophysin projections in the brainstem and spinal cord of normal and homozygous Brattleboro rats. In: *Brattleboro rat* Ann NY Acad Sci 394:759–763
Nürnberger F, Korf HW (1981) Oxytocin- and vasopressin-immunoreactive nerve fibers in the pineal gland of the hedgehog (*Erinaceus europaeus L*). Cell Tissue Res 220:87–97
Nussey SS, Ang VTY, Jenkins JS, Chowdrey HS, Bisset GW (1984) Brattleboro rat adrenal contains vasopressin. Nature 310:64–66
Oertel W, Tappaz ML, Weindl A (1983) Glutamic acid decarboxylase immunoreactive terminals in the circumventricular organs of the rat. Neurosci Lett [Suppl] 14:S266
Okamura H, Fukui K, Koyama E, Tsutou HLO, Tsutou T, Terubayashi H, Fujisawa H, Ibata Y (1983) Time of vasopressin neuron origin in the mouse hypothalamus: examination by combined technique of immunocytochemistry and ^{3}[H]-thymidine autoradiography. Dev Brain Res 9:223–226
Olpe HR, Baltzer V (1981) Vasopressin activates noradrenergic neurons in the rat locus coeruleus: A microiontophoretic investigation. Eur J Pharmacol 73:377–378
Ono TH, Nishimo H, Sasaka K, Muramoto K, Yano I, Simpson A (1978) Paraventricular nucleus connections to spinal cord and pituitary. Neurosci Lett 10:141–146
Palkovits M (1978) Topography of chemically identified neurons in the central nervous system: A review. Acta Morphol Acad Sci Hung 26:211–290
Palkovits M (1979) Microchemistry of microdissected hypothalamic nuclear areas. Int Rev Cytol 56:315–339
Palkovits M (1982) Recent data on neuropeptide mapping in the central nervous system. In: McKerns KG, Pantić V (eds) Hormonally active peptides. Plenum, New York pp 279–306
Paulin C, Dubois PM, Czernichow P, Dubois MP (1978) Immunocytochemical evidence for oxytocin neurons in the human fetal hypothalamus. Cell Tissue Res 188:259–264

Pavel S, Dimitru I, Klepsh I, Dorescu M (1973) A gonadotropin inhibiting principle in the pineal of human fetuses. Evidence for its identity with vasotocin. Neuroendocrinology 13:41–46

Pearlmutter AF, Costantini MG, Loeser B (1983) Characterization of ^{3}H-AVP binding sites in particulate preparations of rat brain. Peptides 4:335–341

Perlow MJ, Reppert SM, Artman HA, Fisher DA, Seif SM, Robinson AG (1982) Oxytocin, vasopressin and estrogen-stimulated neurophysin: Daily patterns of concentration in cerebrospinal fluid. Science 216:1416–1418

Peterson RP (1966) Magnocellular neurosecretory centers in the rat hypothalamus. J Comp Neurol 128:181–190

Pévet P, Reinharz AC, Dogterom J (1980) Neurophysins, vasopressin and oxytocin in the bovine pineal gland. Neurosci Lett 16:301–306

Pfeifer WD, Bookin HB (1978) Vasopressin antagonizes retrograde amnesia in rats following electroconvulsive shock. Pharmacol Biochem Behav 9:261–263

Phillips ML, Weyhenmeyer J, Felix J, Ganten D, Hoffman WE (1979) Evidence for an endogenous brain renin-angiotensin system. Fed Proc 38:2260–2266

Pickard GE (1982) The afferent connections of the suprachiasmatic nucleus of the golden hamster with emphasis on the retinohypothalamic projection. J Comp Neurol 211:65–83

Pickering BT (1976) The molecules of neurosecretion: Their formation, transport and release. Progr Brain Res 45:161–179

Pickering BT (1984) Precursors and products in the formation of neurohypophyseal hormones. Proceed VIIth Int Congr Endocrinol Quebec July 1–7, Elsevier, Amsterdam, p 144

Piekut DT (1983) Ultrastructural characteristics of vasopressin-containing neurons in the paraventricular nucleus of the hypothalamus. Cell Tissue Res 234:125–134

Pittman QJ, Blume HW, Renaud LP (1978) Electrophysiological indications that individual hypothalamic neurons innervate both median eminence and neurohypophysis. Brain Res 157:364–368

Pittman QJ, Blume HW, Renaud LP (1981) Connections of the hypothalamic paraventricular nucleus with the neurohypophysis, median eminence, amygdala, lateral septum and midbrain periaqueductal gray: An electrophysiological study in the rat. Brain Res 215:15–28

Pittman QJ, Lawrence D, McLean L (1982) Central effects of arginine vasopressin on blood pressure in rats. Endocrinol 110:1058–1060

Ramaekers F, Rigter H, Leonard BE (1977) Parallel changes in behaviour and hippocampal serotonin metabolism in rats following treatment with desglycinamide lysine vasopressin. Brain Res 120:485–492

Reaves TA jr, Hayward JN (1979) Immunocytochemical identification of vasopressinergic and oxytocinergic neurons in the hypothalamus of the cat. Cell Tissue Res 196:117–122

Reinharz AC, Vallotton MB (1977) Presence of two neurophysins in the human pineal gland. Endocrinology 100:994–1001

Reinharz AC, Czernichow P, Vallotton MB (1974) Neurophysin-like protein in bovine pineal gland. J Endocrinology 62:35–44

Remy C, Girardie J (1980) Anatomical organisation of two vasopressin-neurophysin-like neurosecretory cells throughout the central nervous system of the migratory locust. Gen Comp Endocrinol 40:27–35

Renaud LP, Rogers J, Sgro S (1983) Terminal degeneration in supraoptic nucleus following subfornical organ lesions: ultrastructural observations in the rat. Brain Res 275:365–368

Reppert SM, Artman HG, Swaminathan S, Fisher DA (1981) Vasopressin exhibits a rhythmic daily pattern in cerebrospinal fluid but not in blood. Science 213:1256–1257

Rexed B (1952) The cytoarchitectonic organization of the spinal cord in the cat. J Comp Neurol 96:415–495

Rhodes CH, Morrell JI, Pfaff DW (1981) Distribution of estrogen-concentrating, neurophysin-containing magnocellular neurons in the rat hypothalamus as demonstrated by a technique combining steroid autoradiography and immunohistology in the same tissue. Neuroendocrinology 33:18–23

Ricardo JA, Koh ET (1978) Anatomical evidence of direct projections from the nucleus of the solitary tract to the hypothalamus, amygdala, and other forebrain structures in the rat. Brain Res 153:1–26

Rigter H, Rijk H, Crabbe JC (1980) Tolerance to ethanol and severity of withdrawal in mice are enhanced by a vasopressin fragment. Eur J Pharmacol 64:53–68

Ritter S, McGlone JY, Kelley KW (1980) Absence of lithium-induced taste aversion after area postrema lesion. Brain Res 201:501–506

Robinson ICAF, Jones PM (1982) Neurohypophyseal peptides in cerebrospinal fluid: recent studies. In: Baertschi AJ, Dreifuss JJ (eds) Vasopressin, corticoliberin and opiomelanocortins. Academic Press, New York, pp 21–32

Rossier J, Battenberg E, Pittman Q, Bayon A, Koda L, Miller R, Guillemin R, Bloom F (1979) Hypothalamic enkephalin neurones may regulate the neurohypophysis. Nature 277:653–655

Rossor MN, Iversen LL, Hawthorn J, Ang VTY, Jenkins JS (1981) Extrahypothalamic vasopressin in human brain. Brain Res 214:349–355

Rossor MN, Hunt SP, Iversen LL, Bannister R, Hawthorn J, Ang VTY, Jenkins JS (1982) Extrahypothalamic vasopressin is unchanged in Parkinson's disease and Huntington's disease. Brain Res 253:341–343

Roth KA, Weber E, Barchas ID (1982) Immunoreactive corticotropin releasing factor (CRF) and vasopressin are localized in a subpopulation of the immunoreactive vasopressin cells in the paraventricular nucleus of the hypothalamus. Life Sci 31:1857–1860

Rubin BS, Menniti VPS, Bridges RS (1983) Intracerebroventricular administration of oxytocin and maternal behavior in rats after prolonged and acute steroid pretreatment. Horm Behav 17:45–53

Saghal A (1984) A critique of the vasopressin-memory hypothesis. Psychopharmacology 83:215–218

Saper CB, Loewy AD, Swanson LW, Cowan WM (1976) Direct hypothalamoautonomic connections. Brain Res 117:305–312

Sar M, Stumpf WE, Miller RJ, Chang KJ, Cuatrecasas P (1978) Immunohistochemical localization of enkephalin in rat brain and spinal cord. J Comp Neurol 182:17–38

Sawchenko PE (1982) Anatomic relationships between the paraventricular nucleus of the hypothalamus and visceral regulatory mechanisms. Implications for the control of feeding bahavior. In: Hoebel BG, Nobin D (eds) Neural basis of feeding and reward. Haer Inst, Brunswick, pp 259:274

Sawchenko PE, Swanson LW (1981) A method for tracing biochemically defined pathways in the central nervous system using combined fluorescence retrograde transport and immunohistochemical techniques. Brain Res 210:31–51

Sawchenko PE, Swanson LW (1982a) Immunohistochemical identification of neurons in the paraventricular nucleus of the hypothalamus that project to the medulla or to the spinal cord in the rat. J Comp Neurol 205:260–272

Sawchenko PE, Swanson LW (1982b) The organization of noradrenergic pathways from the brainstem to the paraventricular and supraoptic nuclei in the rat. Brain Res Rev 4:275–325

Sawchenko PE, Swanson LW (1983) The organization and biochemical specificity of afferent projections to the paraventricular and supraoptic nuclei. Prog Brain Res 60:19–29

Sawchenko PE, Swanson LW (1984) Relationship of oxytocin pathways to the control of neuroendocrine and autonomic function. J Steroid Biochem 20:1500

Sawchenko PE, Swanson LW, Joseph SA (1982) The distribution and cells of origin of ACTH (1–39)-stained varicosities in the paraventricular and supraoptic nuclei. Brain Res 232:365–374

Sawchenko PE, Swanson LW, Vale WW (1984) Corticotropin-releasing factor: Co-expression within distinct subsets of oxtocin-, vasopressin and neurotensin-immunoreactive neurons in the hypothalamus of the male rat. J Neurosci 4:1118–1129

Scharrer B (1967) The neurosecretory neuron in neuroendocrine regulatory mechanisms. Am Zool 7:161–169

Scharrer E (1934a) Stammt alles Kolloid im Zwischenhirn aus der Hypophyse? Frankf Z Pathol 47:134–142

Scharrer E (1934b) Über die Beteiligung des Zellkerns an sekretorischen Vorgängen in Nervenzellen: Frankf Z Pathol 47:143–151
Scharrer E, Scharrer B (1939) Secretory cells within the hypothalamus. In: The hypothalamus. Res Publ Assoc Nerv Ment Dis 20:170–194
Schimchowitsch S, Stoeckel ME, Klein MJ, Garaud JC, Schmitt G, Porte A (1983a) Oxytocin-immunoreactive nerve fibers in the pars intermedia of the pituitary in the rabbit and hare. Cell Tissue Res 228:255–263
Schimchowitsch S, Stoeckel ME, Vigny A, Porte A (1983b) Oxytocinergic neurons with tyrosine hydroxylase-like immunoreactivity in the paraventricular nucleus of the rabbit hypothalamus. Neurosci Lett 43:55–59
Schneider DR, Felt BT, Goldman H (1982) Desglylcyl-8-arginine vasopressin affects regional mouse brain cyclic AMP content. Pharmacol Biochem Behav 16:139–143
Schreibmann MP, Halpern LR (1980) The demonstration of neurophysin and arginine vasotocin by immunocytochemical methods in the brain and pituitary gland of the platyfish, *Xiphophorus maculatus*. Gen Comp Endocrinol 40:1–7
Schrell U, Sofroniew MV, Weindl A, Wetzstein R (1983) Analysis of vasopressin projections from the suprachiasmatic nucleus using combined tracer and peptide immunohistochemistry. Neurosci Lett [Suppl] 14:334
Schubert F, George JM, Rao MB (1981) Vasopressin and oxytocin content of human fetal brain at different stages of gestation. Brain Res 213:111–117
Schultz WJ, Brownfield MS, Kozlowski GP (1977) The hypothalamochoroidal tract. II. Ultrastructural response of the choroid plexus to vasopressin. Cell Tissue Res 178:129–141
Seif SM, Robinson AGH, Zimmerman EA, Wilkins J (1978) Plasma neurophysin and vasopressin in the rat: Response to adrenalectomy and steroid replacement. Endocrinology 103:1009–1015
Sequeira RP, Chaiken IM (1984) Occurence of oxytocin, vasopressin and neurophysins in peripheral nerves. Proceed VIIth Int Congr Endocrinol Quebec, July 1–7, Elsevier, Amsterdam, p 1320
Seybold V, Elde R, Hökfelt T (1981) Terminals of reserpine-sensitive vasopressin-neurophysin neurons in the external layer of the rat median eminence. Endocrinology 108:1803–1809
Sherlock DA, Field PM, Raisman G (1975) Retrograde transport of horseradish peroxidase in the magnocellular neurosecretory system of the rat. Brain Res 88:403–414
Silverman AJ, Zimmerman EA (1975) Ultrastructural immunocytochemical localization of neurophysin and vasopressin in the median eminence and posterior pituitary of the guinea pig. Cell Tissue Res 159:291–301
Silverman AJ, Zimmerman EA (1982) Adrenalectomy increases sprouting in a peptidergic neurosecretory system. Neuroscience 7:2705–2714
Silverman AJ, Zimmerman EA (1983) Magnocellular neurosecretory system. Annu Rev Neurosci 6:357–380
Silverman AJ, Gadde CA, Zimmerman EA (1980a) Effects of adrenalectomy on the incorporation of ^{3}H-cytidine in neurophysin and vasopressin-containing neurons of the rat hypothalamus. Neuroendocrinology 30:285–290
Silverman AJ, Goldstein R, Gadde CA (1980b) The ontogenesis of neurophysin-containing neurons in the mouse hypothalamus. Peptides 1:27–44
Silverman AJ, Hoffman D, Gadde CA, Krey LC, Zimmerman EA (1981a) Adrenal steroid inhibition of the vasopressin-neurophysin neurosecretory system of the median eminence of the rat. Differential effects of corticosterone and desoxycorticosterone administration after adrenalectomy. Neuroendocrinology 32:129–133
Silverman AJ, Hoffman DL, Zimmerman EA (1981b) The descending afferent connections of the paraventricular nucleus of the hypothalamus. Brain Res Bull 6:47–61
Sims K, Hoffman DL, Said SI, Zimmerman EA (1980) Vasoactive intestinal polypeptide (VIP) in mouse and rat brain: an immunocytochemical study. Brain Res 186:165–183
Sladek CD, Gash DM, Khachaturian H, Scott DE, Sladek JR jr (1980) Maturation of the supraoptic nucleus: A multidisciplinary analysis. Peptides 1 [Suppl 1]:51–67
Sladek JR Jr, Sladek CD (1983) Anatomical reciprocity between magnocellular peptides and noradrenaline in putative cardiovascular pathways. In: The neurohypophysis: Structure, function and control. Prog Brain Res 60:437–443

Södersten P, Henning M, Melin P, Ludin S (1983) Vasopressin alters female sexual behaviour by acting on the brain independently of alterations in blood pressure. Nature 301:608–610

Sofroniew MV (1980) Projections from vasopressin, oxytocin and neurophysin neurons to neural targets in the rat and human. J Histochem Cytochem 28:475–478

Sofroniew (1982) Vascular and neural projections of hypothalamic neurons producing neurohypophyseal or ACTH-related peptides. In: Baertschi AJ, Dreifuss JJ (eds) Neuroendocrinology of vasopressin, corticoliberin and opiomelanocortins. Academic Press, New York, pp 73–83

Sofroniew MV (1983 a) Morphology of vasopressin and oxytocin neurons and their central and vascular projections. In: The neurohypophysis: Structure, function and control. Prog Brain Res 60:101–114

Sofroniew MV (1983 b) Vasopressin and oxytocin in the mammalian brain and spinal cord. Trends Neurosci 6:467–472

Sofroniew MV, Glasmann W (1981) Golgi-like immunoperoxidase staining of hypothalamic magnocellular vasopressin, oxytocin and neurophysin neurons in the rat. Neuroscience 6:619–643

Sofroniew MV, Schrell U (1981) Evidence for a direct projection from oxytocin and vasopressin neurons in the hypothalamic paraventricular nucleus to the medulla oblongata: immunohistochemical visualization of both the horseradish peroxidase transported and the peptide produced by the same neurons. Neurosci Lett 22:211

Sofroniew MV, Weindl A (1978) Projections from the parvocellular vasopressin and neurophysin-containing neurons of the suprachiasmatic nucleus. Amer J Anat 153:391–430

Sofroniew MV, Weindl A (1980) Identification of parvocellular vasopressin and neurophysin neurons in the suprachiasmatic nucleus of a variety of mammals including primates. J Comp Neurol 193:659–675

Sofroniew MV, Weindl A (1981) Central nervous system distribution of vasopressin, oxytocin and neurophysin. In: Martinez JL, Jensen RA, Messing RB, Rigter H, McGaugh JL, eds. Endogenous peptides and learning and memory processes. Acad Press New York pp 327–369

Sofroniew MV, Weindl A (1982) Neuroanatomical organization and connections of the suprachiasmatic nucleus. In: Aschoff J, Daan S, Groos G (eds) Springer Berlin, Heidelberg, New York, pp 75–86

Sofroniew MV, Weindl A, Wetzstein R (1977) Immunoperoxidase staining of vasopressin in the rat median eminence following adrenalectomy and steroid substitution. Acta Endocrinol [Suppl] 212:72

Sofroniew MV, Weindl A, Schinko I, Wetzstein R (1979) The distribution of vasopressin-, oxytocin- and neurophysin producing neurons in the guinea pig brain. I. The classical hypothalamo-neurohypophyseal system. Cell Tissue Res 196:367–384

Sofroniew MV, Weindl A, Schrell U, Wetzstein R (1981) Immunohistochemistry of vasopressin, oxytocin and neurophysin in the hypothalamus and extrahypothalamic regions of the human and primate brains. Acta Histochemica [Suppl] 24:79–95

Sofroniew MV, Eckenstein F, Schrell U, Cuello AC (1983) Evidence for colocalization of neuroactive substances in hypothalamic neurons. In: Chan-Palay V, Palay SL (eds) Coexistence of neuroactive substances in neurons. Wiley, New York, pp 73–90

Speidel CC (1917) Gland-cells of internal secretion in the spinal cord of the skates. Thesis, Princeton University, USA

Sterba G (1974) Ascending neurosecretory pathways of the peptidergic type. In: Knowles F, Vollrath L (eds) Neurosecretion – the final neuroendocrine pathway. Springer, Berlin Heidelberg New York, pp 38–47

Sternberger LA, Hardy PH jr, Cuculis JJ, Meyer HG (1970) The unlabeled antibody enzyme method of immunohistochemistry. Preparation and properties of soluble antigen – antibody complex (horseradish peroxidase-antihorseradish peroxidase) and its use in identification of spirochetes. J Histochem Cytochem 18:315–333

Stillman MA, Recht LD, Rosario SL, Seif SM, Robinson AG, Zimmerman EA (1977) The effects of adrenalectomy and glucocorticoid replacement on vasopressin-neurophysin in the zona externa of the median eminence of the rat. Endocrinol 101:42–49

Summy-Long JY, Keil LC, Severs WB (1978) Identification of vasopressin in the subfornical organ region: Effects of dehydration. Brain Res 140:241–250
Sundquist J, Forsling ML, Olsson JE, Åkerlund M (1983) Cerebrospinal fluid arginine vasopressin in degenerative disorders and other neurological diseases. J Neurol Psychiatry 46:14–17
Swanson LW (1977) Immunohistochemical evidence for a neurophysin-containing autonomic pathway arising in the paraventricular nucleus of the hypothalamus. Brain Res 128:346–353
Swanson LW, Hartman BK (1980) Biochemical specificity in central pathways related to peripheral and intracerebral homeostatic function. Neurosci Lett 16:55–60
Swanson LW, Kuypers HGJM (1980) The paraventricular nucleus of the hypothalamus: Cytoarchitectonic subdivisions and organization of projections to the pituitary, dorsal vagal complex, and spinal cord as demonstrated by retrograde fluorescence double-labeling methods. J Comp Neurol 194:555–570
Swanson LW, McKellar S (1979) The distribution of oxytocin- and neurophysin-stained fibers in the spinal cord of the monkey. J Comp Neurol 188:87–106
Swanson LW, Mogenson GJ (1981) Neural mechanism for the functional coupling of autonomic, endocrine and somatomotor responses in adaptive behavior. Brain Res Rev 3:1–34
Swanson LW, Sawchenko PE (1980) Paraventricular nucleus: A site for the integration of neuroendocrine and autonomic mechanisms. Neuroendocrinology 31:410–417
Swanson LW, Sawchenko PE (1983) Hypothalamic integration: organization of the paraventricular and supraoptic nuclei. Annu Rev Neurosci 6:269–324
Swanson LW, Sawchenko PE, Wiegand SJ, Price JL (1980) Separate neurons in the paraventricular nucleus project to the median eminence and to the medulla or spinal cord. Brain Res 198:190–195
Swanson LW, Sawchenko PE, Bérod A, Hartmann BK, Helle KB, Vanorden DE (1981) An immunohistochemical study of the organization of catecholaminergic cells and terminal fields in the paraventricular and supraoptic nuclei of the hypothalamus. J Comp Neurol 196:271–285
Tager H, Hohenboken M, Markese G, Dinerstein RJ (1980) Identification and localization of glucagon-related peptides in rat brain. Proc Natl Acad Sci USA 77:6229–6233
Tanaka M, Versteeg DHG, De Wied D (1977) Regional effects of vasopressin on rat brain catecholamine metabolism. Neurosci Lett 4:321–325
Tigges J, Shantha TR (1969) A stereotaxic brain atlas of the tree shrew (Tupaia glis). Williams and Wilkins, Baltimore
Tribollet E, Dreifuss JJ (1981) Localization of neurones projecting to the hypothalamic paraventricular nucleus area of the rat: a horseradish peroxidase study. Neurosci 6:1315–1328
Ueda S, Kawata M, Sano Y (1983) Identification of serotonin- and vasopressin immunoreactivities in the suprachiasmatic nucleus of four mammalian species. Cell Tissue Res 234:237–248
Van den Dungen HM, Buijs RM, Pool CW, Terlou M (1982) The distribution of vasotocin and isotocin in the brain of the rainbow trout. J Comp Neurol 212:146–157
Van den Pol A (1982) The magnocellular and parvocellular paraventricular nucleus of the rat: intrinsic organization. J Comp Neurol 206:317–345
Vanderhaeghen JJ, Lotstra F, De Mey J, Gilles C (1980) Immunohistochemical localization of cholecystokinin- and gastrin-like peptide in the brain and hypophysis of the rat. Proc Natl Acad Sci USA 77:1190–1194
Vanderhaeghen JJ, Lotstra F, Vandesande F, Dierickx K (1981) Coexistence of cholecystokinin and oxytocin-neurophysin in some magnocellular hypothalamo-hypophyseal neurons. Cell Tissue Res 221:227–231
Vanderhaeghen JJ, Lotstra F, Liston DR, Rossier J (1983) Proenkephalin, [met]enkephalin, and oxytocin immunoreactivities are colocalized in bovine hypothalamic magnocellular neurons. Proc Natl Acad Sci USA 80:5139–5143
Van der Kooy D (1984) Area postrema: site where cholecystokinin acts to decrease food intake. Brain Res 295:345–347

Van der Kooy D, Koda LY (1983) Organization of the projections of a circumventricular organ. The area postrema in the rat. J Comp Neurol 219:328–338
Vandesande F, Dierickx K (1975) Identification of the vasopressin producing and of the oxytocin producing neurons in the hypothalamic magnocellular neurosecretory system of the rat. Cell Tissue Res 164:153-162
Vandesande F, De Mey J, Dierickx K (1974) Identification of neurophysin producing cells. I. The origin of the neurophysin-like substance-containing nerve fibres of the external region of the median eminence of the rat. Cell Tissue Res 151:157–200
Vandesande F, Dierickx K, De Mey J (1975) Identification of the vasopressin-neurophysin producing neurons of the rat suprachiasmatic nuclei. Cell Tissue Res 156:377–380
Vandesande F, Dierickx K, De Mey J (1977) The origin of the vasopressinergic and oxytocinergic fibres of the external region of the median eminence of the rat hypophysis. Cell Tiss Res 180:443–452
Van Leeuwen FW, Caffé R (1983) Vasopressin-immunoreactive cell bodies in the bed nucleus of the stria terminalis of the rat. Cell Tissue Res 228:525–534
Van Leeuwen FW, De Vries GJ (1983) Enkephalin-glial interaction and its consequence for vasopressin and oxytocin release from the rat neural lobe. In: The neurohypophysis: Structure, function, control, Prog Brain Res 60:343–351
Van Leeuwen FW, Wolters P (1983) Light microscopic autoradiographic localization of ^{3}H-arginine-vasopressin binding sites in the rat brain. Neuroscience 41:61–66
Van Leeuwen FW, Swaab DF, De Raay C (1978) Immunoelectron microscopic localization of vasopressin in the rat suprachiasmatic nucleus. Cell Tissue Res 193:1–14
Van Leeuwen FW, Pool CW, Sluiter AA (1983) Enkephalin immunoreactivity in synaptoid elements on glial cells in the rat neural lobe. Neuroscience 8:229–241
Van Ree JM, De Wied D (1977a) Heroin self-administration is under control of vasopressin. Life Sci 21:315–320
Van Ree JM, De Wied D (1977b) Modulation of heroin self-administration by neurohypophyseal principles. Eur J Pharmacol 43:199–202
Van Vossel-Daeninck J, Dierickx K, Van Vossel A, Vandesande F (1979) Electron microscopic immunocytochemical demonstration of separate vasotocinergic and mesotocinergic nerve fibers in the median eminence of the frog hypophysis. Cell Tissue Res 204:29–36
Van Wimersma Greidanus TB (1984) Behavioral effects of neurohypophyseal hormones. Proceed VIIth Int Congr Endocrinol, Quebec July 1–7, Elsevier, Amsterdam, p 30
Van Wimersma Greidanus TB, Dogterom J, De Wied D (1975) Intraventricular administration of anti-vasopressin serum inhibits memory consolidation in rats. Life Sci 16:637–644
Van Wimersma Greidanus TB, De Wied D (1976) Dorsal hippocampus: a site of action of neuropeptides on avoidance behavior? The Neuropeptides. Pharmacol Biochem Behav 5:Suppl 1, 29–33
Van Wimersma Greidanus TB, Bohus B, Kovács GL, Versteeg DHG, Burbach JPH, De Wied D (1983) Sites of behavioral and neurochemical action of ACTH-like peptides and neurohypophyseal hormones. Neurosci Biobehav Rev 7:453–463
Veenstra JA, Romberg-Privee HM, Schooneveld H (1984) Immunocytochemical localization of peptidergic cells in the neuro-endocrine system of the Colorado potato beetle, *Leptinotarsa decemlineata,* with antisera against vasopressin, vasotocin and oxytocin. Histochemistry 81:29–34
Versteeg DHG, Tanaka M, De Kloet ER (1978) Catecholamine concentration and turnover in discrete regions of the brain of the homozygous Brattleboro rat deficient in vasopressin. Endocrinology 103:1654–1661 (1978)
Versteeg DHG, De Kloet ER, Van Wimersma Greidanus TB, De Wied D (1979) Vasopressin modulates the activity of catecholamine containing neurons in specific brain regions. Neurosci Lett 11:69–73
Vigier J, Portalier P (1979) Efferent projections of the area postrema demonstrated by autoradiography. Arch Ital Biol 117:308–324
Vigier J, Rouvière A (1979) Afferent and efferent connections of the area postrema demonstrated by the horseradish peroxidase method. Arch Ital Biol 117:325–339

Voorn P, Buijs RM (1983) An immuno-electronmicroscopical study comparing vasopressin, oxytocin, substance P and enkephalin containing nerve terminals in the nucleus of the solitary tract of the rat. Brain Res 270:169–173
Vorherr H, Bradbury, Leschelle MWB, Hoghoughi M, Kleeman CR (1968) Antidiuretic hormone in cerebrospinal fluid during endogenous and exogenous changes in its blood level. Endocrinology 83:246–250
Wali FA (1984) Effects of oxytocin and vasopressin on ganglionic transmission at the rabbit superior cervical ganglion. Pharmacol Res Commun 16:55–62
Wathes DC, Swann RW (1982) Is oxytocin an ovarian hormone? Nature 297:225–227
Wathes DC, Swann RW, Pickering BT, Porter DG, Hull MGR, Drife O (1982) Neurohypophysial hormones in the human ovary. Lancet 11:410–412
Wathes DC, Swann RW, Birkett SD, Porter DG, Pickering BT (1983) Characterization of oxytocin, vasopressin and neurophysin from the bovine corpus luteum. Endocrinology 113:693–698
Wathes DC, Swann RW, Pickering BT (1984) Variations in oxytocin, vasopressin and neurophysin concentration in the bovine ovary during the oestrus cycle and pregnancy. J Reprod Fertil 71:551–557
Watkins WB (1975) Immunohistochemical demonstration of neurophysin in the hypothalamo-neurohypophyseal system. Int Rev Cytol 41:241–284
Watkins WB (1983) Immunohistochemical localization of neurophysin and oxytocin in the sheep corpora lutea. Neuropeptides 7:51–54
Watkins WB, Choy VJ (1979) Maturation of the hypothalamo-neurohypophyseal system. II. Neurophysin, vasopressin, and oxytocin in the median eminence of the developing rat brain. Cell Tissue Res 197:337–346
Watkins WB, Choy VJ (1980) The impact of aging on neuronal morphology in the rat hypothalamo-neurohypophysial system: an immunohistochemical study. Peptides 1 [Suppl 1]:239–245
Watson SJ, Akil H, Fischli W, Goldstein A, Zimmerman E, Nilaver G, van Wimersma Greidanus TB (1982a) Dynorphin and vasopressin: Common localization in magnocellular neurons. Science 216:85–87
Watson SJ, Seidah NG, Chrétien M (1982b) The carboxy terminus of the precursor to vasopressin and neurophysin: Immunocytochemistry in rat brain. Science 217:833–855
Weber E, Evans CJ, Barchas JD (1982a) Predominance of the aminoterminal octapeptide fragment of dynorphin in the rat brain regions. Nature 299:77–79
Weber E, Roth KA, Barchas JD (1982b) Immunohistochemical distribution of alpha-neoendorphin/dynorphin neuronal systems in rat brain: evidence for colocalization. Proc Natl Acad Sci USA 79:3062–3066
Weber E, Geis R, Voigt KH, Barchas JD (1983) Levels of pro-neoendorphin/dynorphin-derived peptides in the hypothalamo-posterior pituitary system of male and female Brattleboro rats. Brain Res 260:166–171
Weindl A (1973) Neuroendocrine aspects of circumventricular organs. In: Ganong WF, Martini L (eds) Frontiers in Neuroendocrinology. Oxford Univ Press New York, pp 3–32
Weindl A (1983) The blood-brain barrier and its role in the control of circulating hormone effects on the brain. In: Ganten D, Pfaff D (eds) Central cardiovascular control. Springer, Berlin Heidelberg New York Tokyo, pp 151–186 (Current topics in neuroendocrinology 3)
Weindl A, Sofroniew MV (1978) Neurohormones and circumventricular organs. In: Scott DE, Kozlowski GP, Weindl A (eds) Brain-endocrine interaction III. Neural hormones and reproduction. Karger, Basel, pp 117–137
Weindl A, Sofroniew MV (1980) Immunohistochemical localization of hypothalamic peptide hormones in neural target areas. In: Wuttke W, Weindl A, Voigt KH, Driess RR (eds) Brain and pituitary peptides. Karger, Basel pp 97–109
Weindl A, Sofroniew MV (1981) Relation of neuropeptides to mammalian circumventricular organs. In: Martin JB, Reichlin S, Bick KL (eds) Neurosecretion and brain peptides. Raven, New York, pp 303–320

Weindl A, Sofroniew MV (1982) Peptide neurohormones and circumventricular organs in the pigeon. Front Horm Res 9:88–104

Weindl A, Sofroniew MV, Mestres P, Wetzstein R (1980) Immunohistochemische Lokalisation von neurohypophysären Peptiden im Gehirn der Taube (*Columba livia*). Anat Anz 74:769–774

Weindl A, Sofroniew MV, Fuchs E, Wetzstein R (1981) The distribution of neurohypophyseal peptides in the brain of the tree shrew (*Tupaia belangeri*). Folia Morphol (Warsz) 29:41–45

Weindl A, Bruhn T, Parvizi N, Ellendorff F (1984) Ontogeny of neurohypophyseal peptides containing neurons in the pig brain. In: Ellendorff F, Gluckman P, Parvizi U (eds) Fetal neuroendocrinology. Perinatology Press New York, pp 35–38

Weingartner H, Gold P, Ballenger JC, Smallberg SA, Summers R, Rubinow DR, Post RM, Goodwin FK (1981) Effects of vasopressin on human memory functions. Science 211:601–603

Whitnall MH, Gainer H, Cox BM, Molineaux CJ (1983) Dynorphin (1–8) is contained within vasopressin neurosecretory vesicles in rat pituitary. Science: 1137–1139

Wiegand SJ, Price JL (1980) The cells of origin of the afferent fibers to the median eminence in the rat. J Comp Neurol 192:12–19

Wittkowski W, Bock R (1972) Electron microscopical studies of the median eminence following interference with the feedback system anterior pituitary-adrenal cortex. In: Brain-endocrine interaction. Median eminence, structure and function. Karger, Basel, pp 171–180

Wolf G, Trautmann B (1977) Ontogeny of the hypothalamo-neurohypophyseal system in rats – an immunohistochemical study. Endokrinologie 19:222–226

Yamamura HI, Gee KW, Brinton RE, Davis TP, Hadley M, Wamsley JK (1983) Light microscopic autoradiographic visualization of (3H)-arginine vasopressin binding sites in the rat brain. Life Sci 32:1919–1924

Yamashita H, Inenaga K, Koizumi K (1984) Possible projections from regions of paraventricular and supraoptic nuclei to the spinal cord: electrophysiological studies. Brain Res 296:373–378

Yulis CR, Rodriguez EM (1982) Neurophysin pathways in the normal and hypophysectomized rat. Cell Tissue Res 227:93–112

Zandberg P, Palkovits M, De Jong W (1977) The area postrema and control of arterial blood pressure; absence of hypertension after excision of the area postrema in rats. Pflügers Archiv 372:169–173

Zerbe RL, Kirtland S, Faden AI, Feuerstein G (1983) Central cardiovascular effects of mammalian neurohypophyseal peptides in conscious rats. Peptides 4:627–630

Zerihun L, Harris M (1981) Electrophysiological identification of neurons of paraventricular nucleus sending axons to both the neurohypophysis and the medulla in the rat. Neuroscience 23:157–160

Zimmerman EA, Carmel PW, Husain MK, Ferin M, Tannenbaum M, Frantz AG, Robinson AG (1973a) Vasopressin and neurophysin: high concentration in monkey hypophyseal portal blood. Science 182:925–927

Zimmerman EA, Hsu KC, Robinson AG, Carmel PW, Frantz AG, Tannenbaum M (1973b) Studies of neurophysin secreting neurons with immunoperoxidase technique employing antibody to bovine neurophysin. I. Light microscopic findings in monkey and bovine tissues. Endocrinology 92:931–940

Zimmerman EA, Krupp L, Hoffman DL, Matthew E and Nilaver G (1980) Exploration of peptidergic pathways in brain by immunocytochemistry: A ten years perspective. Peptides 1 [Suppl 1]:3–10

Acknowledgement. The assistance of Drs. R. Lang, Heidelberg, D. Schams, Weihenstephan, and M. V. Sofroniew, Oxford, who provided antisera for immunohistochemical peptide localization demonstrated in several of the figures of this article is gratefully acknowledged.

Note Added in Proof

Vasopressin perikarya in the bed nucleus of the stria terminalis and medial amygdaloid nucleus disappeared in male rats 15 weeks after castration [De Vries GJ, Buijs RM, Van Leeuwen FW, Caffé AR, Swaab DF (1985) The vasopressinergic innervation of the brain in normal and castrated rats. J Comp Neurol 233:236–254]

Subject Index

A. M. Neville, M. J. O'Hare

The Human Adrenal Cortex

Pathology and Biology - An Integrated Approach

1982. 173 figures. XIV, 354 pages
ISBN 3-540-11085-2

Contents: Introduction. - Historical Aspects. - The Normal Adrenal Cortex. - Hypercorticalism. - Hypocorticalism. - Other Pathological Changes. - Appendix A: Plasma Steroids. - Appendix B: Urinary Steroids. - References. - Subject Index.

Recent advances in our understanding of many pathological processes have been associated in no small measure with improved understanding of the structure and function of normal tissues. Virchow's dictum that all diseases are, in the final analysis, reducible to disturbances in the functional activity of the primary units (i.e. the cells) of normal tissues is one that can still be applied with profit. Nowhere is this better illustrated than in endocrinology, especially in relation to the human adrenal gland and its pathology.

This treatise is a unique guide to the structure and function of the human adrenal cortex. Incorporating many of their own results using monolayer adrenal cell cultures and steroid analysis, the authors have produced a work that illustrates the numerous variable patterns of both the normal and diseased adrenal. In so doing, they pave the way for an improved understanding of adrenocortical pathology. Their textual comments are accompanied by an extensive atlas to provide histopathologists and endocrinologists with reliable reference for both diagnosis and subsequent patient monitoring.

Springer-Verlag
Berlin
Heidelberg
New York
Tokyo

GPSR Compliance

The European Union's (EU) General Product Safety Regulation (GPSR) is a set of rules that requires consumer products to be safe and our obligations to ensure this.

If you have any concerns about our products, you can contact us on ProductSafety@springernature.com

In case Publisher is established outside the EU, the EU authorized representative is:

Springer Nature Customer Service Center GmbH
Europaplatz 3
69115 Heidelberg, Germany

Batch number: 10620881

Printed by Printforce, the Netherlands